BIOCHEMISTRY OF
PHOTOSYNTHESIS

BIOCHEMISTRY OF PHOTOSYNTHESIS
3rd Edition

R. P. F. Gregory
Department of Biochemistry and Molecular Biology
University of Manchester

A Wiley-Interscience Publication

JOHN WILEY & SONS
Chichester · New York · Brisbane · Toronto · Singapore

Library of Congress Cataloging-in-Publication Data:

Gregory, R. P. F.
    Biochemistry of photosynthesis / R.P.F. Gregory.—3rd ed.
      p.   cm.
    'A Wiley–Interscience publication.'
    Bibliography: p.
    Includes indexes.
    ISBN 0 471 91899 7
    1. Photosynthesis. I. Title.
    [DNLM: 1. Photosynthesis.   QK 882 G823b]
  QK882.G83
  581.1'3342—dc19
  DNLM/DLC
  for Library of Congress   88-14428
                                                             CIP

British Library Cataloguing in Publication Data:

Gregory, R. P. F. (Richard Paul Fitzgerald),
  1938–
  Biochemistry of photosynthesis.—3rd ed.
  1. Plants. Photosynthesis
  I. Title
  581.1'3342

  ISBN 0 471 91899 7

Printed and bound in Great Britain by the Bath Press

To Mark, David, Adam and Kate

Contents

Preface to the First Edition

The process of photosynthesis, in which the energy of light (sunlight) is chemically captured by living organisms, and on which the whole of our planet's life-stock depends, is worthy of considerable stress in an undergraduate course of biochemistry, the more so if it provides for a synthesis of topics normally isolated by the necessarily linear nature of such courses. For example, the metabolic pathways of photosynthesis form a useful antithesis in the understanding, at an elementary level, of the pathways of glycolysis and the pentose cycle. The same applies to the cytochromes of the thylakoid, the production of oxygen from water, NADP reduction and photophosphorylation, all of which can be profitably compared with analogous processes in mitochondria. The chloroplast itself, in its relation to the cell, presents a most striking example of biochemical compartmentation. The first aim of this text is to provide an introduction to photosynthesis on the above basis. It is hoped that this introduction, Part I, will be with some selection valuable in courses of botany, and possibly at sixth-form level as well.

Since an introduction involves simplification, which is somewhat unsound, the text continues in Part II towards a second purpose, that of presenting an account of the subject giving the principal points of view (in 1970) and the experimental work and argument by which they were defended. As the writing progressed, however, it became clear that simplification would reappear, and the reader should be aware of two serious manifestations of it. First, the historical aspect and the development of the present-day concepts of photosynthesis, has been largely suppressed, except where old or obsolescent terms and hypotheses may still be unearthed by students and cause confusion. References in the text, and figures reproduced from previous publications, have in the main been selected for their utility and clarity of exposition rather than for evidence of priority. Secondly, the style adopted for the work is that of dividing up the field into a few areas, one to each chapter, and within each chapter to set out sections each written round a discrete point of view, aiming to cover the greater part of the area concerned.

In accord with the aim of supporting a course of lectures in this subject, I have given at the end of Part I a selection of problems, some numerical, some requiring discussion. To enable the most use to be made of Part II as a reference section, the index has been laid out in an extended manner, for which I am grateful to the publishers.

It is a pleasure to thank all my colleagues who have helped with suggestions, particularly Dr I. West and Dr A. G. Lowe. I am indebted to Mr A. Greenwood, Dr H. Bronwen Griffiths, Mr R. Bronchart, Professor R. B. Park and Dr G. Cohen-Bazire for electron micrographs, and Professor D. A. Walker for an autoradiograph. I gratefully acknowledge the painstaking criticism offered by Professor F. R. Whatley, Professor Walker and Dr A. R. Crofts. My thanks are also due to Mrs D. M. Warrior for typing the manuscript.

RICHARD GREGORY
Department of Biological Chemistry,
The University of Manchester.
Spring 1971.

Preface to the Second Edition

It is a privilege to have an opportunity of revising one's work. Apart from correcting errors, which I regret having made and still more my students having seen, it is a pleasure to record the advances made, particularly in the two fields of carbon metabolism and photophosphorylation. The layout and style of the first edition have been preserved. While the text cannot claim to be a review, it is hoped that the student will find some account of most major experiments and discussion up to the end of 1975.

I am most grateful to Professors C. C. Black and A. Staehelin for electron micrographs, to Professors D. O. Hall and D. A. Walker who provided most helpful criticism and gave me access to material in advance of its publication, to Professor V. Massey for Figure 8.1(d), and to Dr M. C. W. Evans for help with Table 8.2.* Dr J. Barber, Dr S. Raps and Dr R. Hill were generous with their time in reading and discussing the manuscripts; but they were in no way responsible for deficiencies remaining. I thank Mrs J. Black for ably typing the manuscript, and Mr H. Mann for his help with the task of indexing.

R. P. F. GREGORY
Manchester
February 1976.

---

* *Publisher's note*: figure and table numbers refer to the second edition.

Preface to the Third Edition

Photosynthesis research has shared in the world-wide growth of biochemical research, and this text records the advances made and the new coherence in the story that they tell.

The object of the text remains substantially the same, to support a course of lectures to advanced undergraduates in the School of Biological Sciences at Manchester University, by describing the experimental methods and arguments through which the advances were achieved, and to provide an indication of the present areas of emphasis and progress. Comparisons have been drawn wherever possible with analogous findings in other areas of biochemical research. Obviously I hope to tempt the discerning student into this field. Secondly, the aim is to assist in providing an advanced overview both to researchers starting out in photosynthesis, and for established workers elsewhere who wish to pick up rapidly individual topics that may be relevant to their own fields. As far as possible the sections of the text are intended to be read independently.

'The Biochemistry of Photosynthesis' is intended to describe the search for an account of the biological phenomenon at the molecular level. It has been necessary to interpret this fairly strictly, on the one hand taking an admittedly simple view of the physical processes of excited states of pigments, and of the primary electron transport processes in photosynthetic reaction centres, and on the other of cutting the account short in the realm of plant physiology where the macroscopic relations of structure and function provide the visible evidence of the importance of photosynthesis in the world around us.

It is a matter for regret that the earlier work in the subject has been considerably suppressed; there are often insights to be gained from the question 'How did we ever get here?' For the older viewpoints, however, we must consult the older books.

A second problem has been the choice of references to be cited in the text. There are both too few and too many. I have provided titles of articles to enable readers to make their own judgment.

I am most grateful to the many colleagues who have sent me reprints, photographic prints and other material, and to colleagues including Dr Tristan A. Dyer, Dr J. Rosamond, Dr K. V. Rowsell and Dr J. P. Thornber, for their time in discussions. I particularly acknowledge the support and forbearance of my wife Kathleen.

RICHARD GREGORY
Department of Biochemistry and Molecular Biology,
University of Manchester.

Abbreviations

δALA	δ-aminolevulinic acid
ACP	acyl-carrier protein
Asp	aspartic acid
BChl	bacteriochlorophyll
BPGA	1,3-bisphospho-D-glycerate
BPh	bacteriophaeophytin
bR	bacteriorhodopsin
CAM	Crassulacean acid metabolism
CC	core complex
CCCP	carbonyl cyanide *m*-chlorophenylhydrazone
Cb.	*Chlorobium*
CD	circular dichroism
Chl	chlorophyll
Chr.	*Chromatium*
Cp.	*Chloropseudomonas*
Cys	cystine
DAD	diaminodurene
DBMIB	2,5-dibromo-3-methyl-6-isopropyl-*p*-benzoquinone
DCPIP	2,6-dichlorophenolindophenol
DDG	diacyldigalactoglycerol
DG	diacylglycerol
DGDG	digalactosyldiacylglycerol
DGG	diacylgalactosylglycerol
DHAP	dihydroxyacetone phosphate
DL	delayed light
DSQ	diacylsulphoquinovosylglycerol
E.	*Escherichia*
E4P	D-erythrose 4-phosphate

ESR	electron spin resonance
F1,6BP	fructose 1,6-bisphosphate
F1,6Base	fructose 1,6-bisphosphatase
F6P	fructose 6-phosphate
F6P kinase	fructose 6-phosphate kinase
FAD	flavin adenine dinucleotide
FBP phosphatase	fructose bisphosphate phosphatase
FMN	flavin mononucleotide
FNR	ferredoxin : NADP reductase
FP	flavoprotein
FQR	ferredoxin : quinone reductase
G1,6BP	glucose 1,6-bisphosphate
G6P	glucose 6-phosphate
GAP	glyceraldehyde 3-phosphate
GAPDH	glyceraldehyde 3-phosphate dehydrogenase
Glu	glutamic acid
HiPIP	high-potential iron protein
His	histidine
HPLC	high-pressure liquid chromatography
hR	halorhodopsin
LDAO	lauryldimethylamine-N-oxide
LHC	light-harvesting complex
S-met	S-methionine
MGDG	monogalactosyldiacylglycerol
MK	menaquinone
MKH$_2$	menaquinol
NEM	N-ethylmaleimide
NMR	nuclear magnetic resonance
NQNO	2-n-nonylhydroxyquinoline N-oxide
OAA	oxaloacetate
3-OH-3-Me-glutaryl CoA	3-hydroxy-3-methyl-glutaryl CoA
P.	Prosthecochloris
PA	phosphatidic acid
PAGE	polyacrylamide gel electrophoresis
PAR	photosynthetically active radiation

PC	phosphatidylcholine
PCB	phycocyanobilin
PDK	pyruvate phosphate dikinase
PEB	phycoerythrobilin
PG	phosphatidylglycerol
PGA	3-phosphoglycerate
PGK	phosphoglycerate kinase
Ph	phaeophytin
PI	phosphatidylinositol
PMF	protonmotive force
PMS	phenazine methosulphate
PQ	plastoquinone
PQH_2	plastoquinol
PRuK	phosphoribulokinase
PSI	photosystem I
PSII	photosystem II
PUB	phycourobilin
PUVB	photobiological ultraviolet range B
Q	quinone
QH_2	quinol
R5P	D-ribose 5-phosphate
R.	*Rhodospirillum*
Rb.	*Rhodobacter*
Rps	*Rhodopseudomonas*
Ru5P	ribulose 5-phosphate
RuBisCO	ribulose bisphosphate carboxylase
RuBP	D-ribulose 1,5-bisphosphate
SDS	sodium dodecyl sulphate
SH1,7BP	D-sedoheptulose 1,7-bisphosphate
SH7P	sedoheptulose 7-phosphate
SHBP aldolase	sedoheptulose bisphosphate aldolase
SHBP phosphatase	sedoheptulose bisphosphate phosphatase
SR	slow rhodopsin
TPB	tetraphenylborate
Tyr	tyrosine
UQ	ubiquinone
UQH_2	ubiquinol
Xu5P	D-xylulose 5-phosphate

Chapter 1

Structure and Function in the Photosynthesising Cell

1.1 Introduction: Life Needs Energy

The biochemist investigates biological problems using the techniques of chemistry. The aim is to provide an account of living things at the molecular level. Clearly there is the risk of losing sight of the organism as a whole or the community in which it lives, but the reward is the particular insight into the world of biology that comes from this kind of 'reductionism'.

Photosynthesis is one of two solutions to a problem faced by all living organisms, the supply of energy. If the energy source is light, the organism is performing photosynthesis, and its lifestyle is phototrophic. Alternatively, the organism may take in materials from its environment that will react to release energy, as in respiration or fermentation, and its lifestyle is chemotrophic. Most plants are phototrophic, fungi and most animals are chemotrophic and bacteria may be either or both.

The fundamental need for an energy supply can be simply stated as the need to preserve order. Order in biological systems is not so easily defined, but it is clear that it exists, from contemplating death. The cell is subject to constant attack by disruptive forces. Macromolecules (protein, nucleic acids, lipids and carbohydrates) are subject to random hydrolysis, and cells that live in air have to manage a continual oxidation of many molecules. The lipid phase, represented by membranes, can re-form perhaps as droplets, and the membranes themselves, which cannot be totally impermeable, cannot avoid some leakage of water and solutes into or out of the cell. Mechanical and other forms of shock cause similar damage. These injuries are all spontaneous, and unchecked they lead at a greater or lesser rate to the loss of the identity of the cell as distinct from its environment. That is to say, spontaneous processes continue until they reach a state of equilibrium.

The constancy of structure of a living cell is only apparent. The damage has to be made good. Irreversible changes cannot be reversed directly, and the damaged structures are broken down, the parts scavenged, and replacements rebuilt. Assembling the structures of the living cell and repairing the continual damage requires an input of energy. In considering the balance of energy, we need to consider not only heat energy (which is measured by the heat of combustion of the substances present) but also an energy-like expression derived from the orderliness of the system. The two together are known as Gibbs energy (commemorating J. Willard Gibbs, 1839 to 1903).

1.1.1 Gibbs energy

Gibbs energy (G) is an algebraic sum, of one term that represents mechanical, electrical or heat energy (H), and a subtractive term TS which is the product of the absolute temperature T and entropy S: thus $G = H - TS$. We are only interested in changes in G in relation to processes where there is a change in H or S; absolute values are not required. For practical purposes T is constant.

Entropy increases with disorder, where examples of disorder are increased dilution, melting or dissolution of solids, evaporation of liquids or breakdown of large molecules into many subunits. Spontaneous processes are characterised by a loss of G (that is, by a negative change in G) either because of loss of heat, as when fuel burns, or because of an increase in disorder, as when salt dissolves in water. It is the usual practice to define for each chemical reaction or process a 'standard Gibbs energy change' (ΔG_0), where 'standard' means calculated for the condition where all reactants and products are in standard states, usually unit concentration. If the concentrations in a given situation are not unity, as obviously must be the case in most biochemical processes, the actual ΔG for the reaction differs from ΔG_0 by a term obtained from the concentration product according to the law of mass action:

$$\Delta G = \Delta G_0 + RT \ln \frac{[\text{products}]}{[\text{reactants}]}$$

where R is the ideal gas constant.

As the reaction proceeds the concentrations change and ΔG diminishes, until the concentration product is equal to the equilibrium constant, and $\Delta G = 0$. Equilibria exist when all possible processes are accompanied by neither an increase nor a decrease in G. In fact ΔG_0 values are usually obtained from the equilibrium constant by means of the relationship

$$\Delta G_0 = -RT \ln K$$

where K is the equilibrium constant.

In an anabolic process such as the synthesis of protein from amino acids, S would decrease and G would increase; synthesis would not happen unless it could be coupled to another process in which a greater counterbalancing loss of G occurred. Such a process is the hydrolysis of ATP: the loss in G accompanying the hydrolysis of ATP outweighs the gain in G accompanying the synthesis of the protein molecule.

1.1.2 The significance of ATP

For the present purpose, the essential feature of ATP is the anhydride link of the terminal phosphate. Hydrolysis of an anhydride liberates heat energy arising chiefly from the resonance stabilisation of the two acid anions, and the total entropy is increased by the increase in the number of molecules. The effect is that the equilibrium constant for the hydrolysis of ATP under the conditions of an average cell is of the order of 10^5 M:

$$([ADP][P_i]/[ATP])_{eq} = 10^5 \text{ M}$$

The cell is, however, provided with means of driving the synthesis of ATP, one of which is photosynthesis, and the concentration ratio can be maintained at approximately 10^{-4} M. The nine orders of magnitude between these two values represents the ability of ATP to act as an energy store in the cell.

Living things survive and grow only because they can maintain their reserves of ATP—'the energy currency of the cell'—so the problem for the cell now becomes that of finding an ATP supply. The chemotrophs couple the formation of ATP to the spontaneous reactions of foodstuffs (fermentation or respiration reactions), and they excrete waste materials which are closer to the environmental equilibrium than was the foodstuff. (Foodstuffs do slowly oxidise, as evidenced by the deterioration in the flavour of deep-frozen meals!) The chemotrophic cell therefore can be said to be catalysing the reaction by which foodstuff turns into waste (e.g. glucose $+ O_2 \rightarrow CO_2 + H_2O$). It achieves this by providing a pathway for the reaction which is faster than the environmental rate (consider the analogy with a hydro-electric power station which provides a fast pathway or pipeline for draining a lake) (Fig. 1.1a).

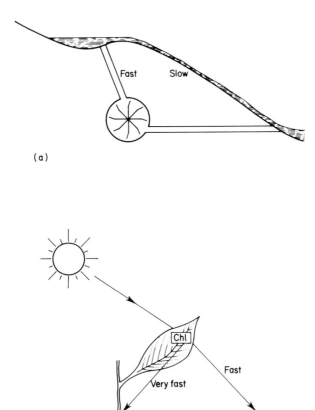

Figure 1.1. Energy conservation. (a) The outflow from the mountain lake down the river represents a non-conserving path for the conversion of gravitational energy to heat. The hydro-electric station represents a much faster pathway, in which some of the gravitational energy can be conserved as useful work. (b) Sunlight is temporarily captured and held by the pigment chlorophyll. Fast processes exist for the energy to be dissipated as heat. Photosynthesis provides a very fast pathway so that some of the captured light energy is conserved as chemical energy

1.1.3 Capturing the energy of light

The phototroph employs a different strategy. The cell sets up a pigment, that is a molecule containing a chromophore group. A chromophore by definition absorbs light. All molecules absorb electromagnetic radiation of some wavelength, ultraviolet, visible or infrared, and the chromophore of the pigment is adapted to absorb in a region where there is little competition from the other materials of the cell. Not all wavelengths are suitable for capturing energy into ATP. Infrared light, for example, with a wavelength range of 1 μm to 1 mm, is absorbed as heat. Shorter wavelengths are necessary, because they give rise to electronically excited states of molecules. Excited states are chemically reactive, if their lifetime is long enough. The shorter the wavelength of the light absorbed, the greater the energy of the corresponding excited state and hence its reactivity (in general). The range of wavelengths that we call visible light (400–800 nm) is close to the lowest energy possible for excited states. Ultraviolet light (<400 nm) is dangerous. UV-absorbing groups occur widely in cells, particularly in proteins and nucleic acids. The high reactivity of the corresponding excited states results in destruction of these essential materials. It is no coincidence that the photosynthetic range of wavelengths is roughly in correspondence with the visible range.

An excited state has a limited lifetime. The phototroph shortens it further by providing a deactivation pathway in which some of the energy of the excitation is conserved, principally in the form of ATP. The phototroph in a sense cata-lyses the warming of the environment by means of the light energy of the sun (Fig. 1.1b). This is superficially similar to the overall catalytic behaviour of the chemotroph, but the difference lies in the fact that chemotrophs can very readily exhaust the environment and jeopardise their own existence by their success. The temperature difference between the earth and sun is, however, fixed, and the supply of energy for phototrophs is inexhaustible. In economic terms, energy from photosynthetic sources is a renewable resource, in the sense that fossil fuels are not.

Chemotrophs rely on the non-equilibrium materials produced by the pho-totrophs, and this is illustrated by Fig. 1.2a, showing the concept of energy flow from the sun through plants to animals, eventually giving rise to environmen-tal heat. Figure 1.2b shows the well known carbon cycle, which should remind us that it is the unique chemistry of carbon that makes life possible, and Fig. 1.2c shows the oxygen cycle, from which it should be noted that the oxygen of the earth's atmosphere is of biological and photosynthetic origin, and is turned over in the surprisingly short period of some 2000 years. The fixation of carbon dioxide, and production of oxygen, is due to the green plants and some pho-tosynthetic bacteria which have autotrophic nutrition: carbon dioxide is their sole source of carbon. Autotrophy is not synonymous with photosynthesis how-ever, as other photosynthetic bacteria, in common with animals, require organic carbon and are known as heterotrophs.

1.1.4 Membranes

The primary role of photosynthesis is to provide a means of generating ATP from light energy. A secondary role (particularly in green plants) is to gen-erate a reduced coenzyme, and an oxidised waste product (oxygen in green plants). The ATP, and reduced coenzyme, provide the driving force for the metabolism of the cell: there is no essential difference between the metabolism of chemotrophs and heterotrophs. A membranous structure is required to fulfil both roles, first, to separate the oxidised and reduced photo-products which would be likely, in a homogeneous system, to react together, and secondly because a membrane-bounded closed vesicle provides a means of accumulating and storing energy in an intermediate form, from which the synthesis of ATP molecules can be achieved.

All kinds of phototrophic system therefore rely on the activity of particular membranes. Biological membranes in general are composed of protein and lipid. Proteins may be either intrinsic or extrinsic. Intrinsic proteins are embedded in the lipid, and extrinsic proteins are attached to intrinsic proteins. Intrinsic proteins may be in contact with the aqueous phase on one or other side, or on both sides, when they are termed 'membrane-spanning proteins'. Some membrane-spanning proteins are able to provide channels by which water and specific polar or ionic solutes cross the membrane. Materials which have an appreciable solubility in non-polar solvents (lipophilic molecules), on the other

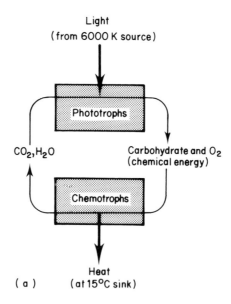

Figure 1.2. Organisms and the environment. Living organisms are dependent on their environment, but the environment also depends on the organisms. (a) The inflow of energy from the sun allows phototrophic organisms to enrich the environment with chemical energy represented in the diagram by carbohydrate and oxygen. Chemotrophs deactivate the chemical energy. Ultimately all the energy is lost as heat. (b) The carbon cycle. The carbon dioxide in the air is maintained and exchanged in a period of a few hundred years by the activity of living organisms. (c) The oxygen cycle. The oxygen of the Earth's atmosphere is maintained in the same way as (b), with a period of a few thousand years

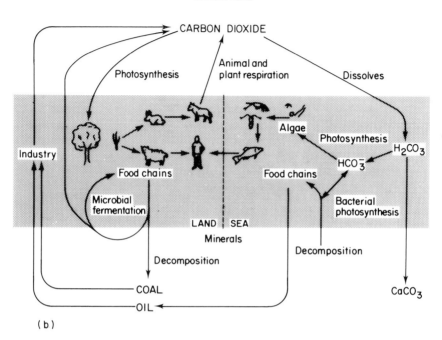

(b)

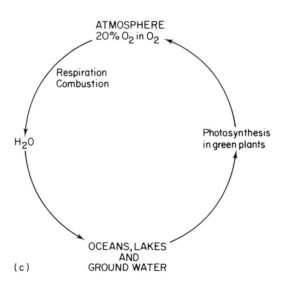

(c)

Figure 1.2. (b, c) (*see legend on p. 6*)

hand, can cross the membrane without the need for protein carriers. Membrane proteins, like proteins in general, are frequently composed of several polypeptide chains (which may or may not be identical), and are then known as complexes. The functions of protein complexes in photosynthetic membranes include supporting arrays of light-absorbing molecules, conducting electrons and acting as enzymes.

The lipid part of the membrane may (in general) contain lipids that form a bimolecular leaflet (see Fig. 1.3) or (in particular in photosynthetic membranes) surround protein complexes as if lubricating their movement in the plane of the membrane (see Fig. 12.8). Small molecules such as plastoquinone, dissolved in the lipid, diffuse rapidly from complex to complex. The membrane, a lipid or non-aqueous phase, is therefore to be considered as much of a working environment as the two aqueous phases that it separates.

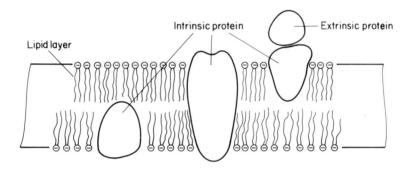

Figure 1.3. Generalised structure of biological membranes. The bimolecular leaflet structure is formed by polar, diacyl lipids, in which intrinsic membrane proteins are found, spanning the membrane thickness or located specifically towards one or the other sides. Extrinsic membrane proteins are attached to intrinsic proteins. Since the intrinsic proteins can move laterally, this structure is known as the 'fluid mosaic', described by S. Jonathan Singer and G. L. Nicolson in 1972

The cell membrane (which separates protoplasm from the surrounding medium) is different on its two sides. The outside surface carries carbohydrate oligo- and polysaccharide structures attached to membrane proteins, while the proteins that are exposed on the inside surface are often found to have phosphate groups attached, usually to serine residues. It is an important part of membrane theory that membrane-spanning proteins retain their orientation, that is, they cannot turn over or 'tumble', although they may rotate about an axis normal to the membrane, change their shape or 'conformation' and diffuse laterally within its plane. It has become a convention (originally used by electron microscopists) to refer to the external side as the E face and the internal side as the P face (for 'protoplasmic').

The same theory applies to membranes within the cell. In each case it is found that one surface is in contact with protein-rich protoplasm and the other with a medium containing few proteins, which can be regarded as the equivalent of the external medium. Hence all biological membranes have E and P surfaces. The interrelationships of the various membranes and their functioning are greatly clarified by making this distinction between the two surfaces.

Photosynthetic membranes and the space they enclose accumulate energy in order to synthesise ATP. This is achieved by the following mechanism, according to the 'chemiosmotic hypothesis' worked out by P. Mitchell and honoured by a Nobel Prize in 1978.

Membranes such as the cell membrane of bacteria, the inner (cristal) membrane of mitochondria and the thylakoid membrane of cyanobacteria and chloroplasts, separate two aqueous regions (on their P and E sides, see above). These aqueous phases differ in pH, and there is also an electrical membrane potential. This state of affairs is maintained by the activity (photosynthetic or respiratory) in the membrane. A proton (H^+) will tend to diffuse towards the higher pH (molecules diffuse down concentration gradients, driven by the increase in entropy resulting from the dilution), and will also release energy because it is a positive charge moving down a voltage gradient. The membrane is, however, impermeable to hydrogen ions which are constrained to use pores in special protein complexes (F-ATPases) that are able to couple the passage of hydrogen ions to the formation of ATP. They are called ATPases because when they are experimentally detached from the membrane they hydrolyse ATP.

The Gibbs energy required to synthesise a mole of ATP by reversal of its hydrolysis may be calculated from the figures given above for the actual cell concentration product (Z, 10^{-4} M) and the equilibrium constant ((K, 10^5 M):

$$\Delta G = -RT \ln (Z/K)$$

For the ratio of approximately 10^9, at $25°C$, $\Delta G = 52$ kJ mol^{-1}. The sign is positive because the system gains energy.

The diffusion of hydrogen ions involves a Gibbs energy change made up of two parts. First, the Gibbs energy released by the diffusion of one mole of a substance from concentration c_1 to concentration c_2 is given by

$$\Delta G = RT \ln (c_2/c_1)$$

and when hydrogen ions are considered,

$$\Delta G = 2.303 \, RT \, (\mathrm{pH}_1 - \mathrm{pH}_2)$$

since pH $= -\log_{10}c$, and the factor 2.303 converts natural to base ten logarithms. Secondly, the Gibbs energy released by a positive ion with one mole of charge moving across a potential difference of V, towards the negative, is given by

$$\Delta G = -FV$$

where F is the Faraday, 96,460 coulombs mol^{-1}. The sign is negative because the system loses energy.

The F-ATPase complex in the membrane requires the passage of (probably) three hydrogen ions to synthesise one molecule of ATP. It will be able to synthesise ATP whenever the total ΔG obtained from summing the contributions of the diffusive and the electrical Gibbs energies amounts to (arithmetically) more than $-52/3$ kJ mol^{-1} H^{+}.

Biological membranes can support potentials of 0.2 V, and (or) some 3.5 units of pH difference. A potential difference of 0.2 V is equivalent to a ΔG of -19 kJ mol^{-1} H^{+}; 3.5 units of ΔpH are equivalent to -20 kJ mol^{-1} H^{+}. This storage of energy is sufficient to support ATP production by the F-ATPase complex, and the process, being common to membranes of chloroplasts, mitochondria and bacterial cells, reveals a satisfying unity of Nature.

1.1.5 Conversion of light energy

There are two totally distinct means by which light energy gives rise to the above-mentioned store of energy: one is based on a single protein, bacterio-rhodopsin, found in *Halobacterium halobium* (a halophilic archaebacterium). Bacteriorhodopsin is an intrinsic protein in the cell membrane, and contains retinal as a prosthetic group. When the retinal absorbs light, it drives a series of changes which result in the protein expelling protons from the protoplasm through the cell membrane into the periplasmic space (effectively an E space), thus generating both a pH difference and a membrane potential.

All other forms of photosynthesis rely on 'reaction centres' which contain chlorophyll (or bacteriochlorophyll) attached to intrinsic proteins in membranes. Absorption of light leads to a photochemical electron transfer reaction forming an oxidised and a reduced product, which are separated into different phases (either of the E or P aqueous phases or the membrane phase). Subsequent reactions of these products (which are essentially homologous with some sections of the electron transport chain of aerobic respiration in mitochondria) have the effect of transferring protons from one side of the membrane to the other, thus setting up the pH difference and membrane potential on which the synthesis of ATP depends.

In the different examples, the vesicle membranes do not always have the same orientation; for example thylakoids have an E phase on the inside, while bacterial cell membranes have a P phase on the inside. Nevertheless the mechanics work the same way: regardless of orientation, in all cases the P side (with respect to the E side) is more alkaline, negatively charged and reducing; also the proteins on the P side of membranes tend to be phosphorylated, and those on the E side glycosylated.

1.2 Photosynthetic Structures

For the present purposes, bacteria (including the blue-green bacteria or algae) have a cell type known as prokaryotic, in which the cell itself is the only protoplasmic compartment. There are no mitochondria, chloroplasts or nuclei. The cell membrane is bounded by the periplasmic membrane, which is the inside surface of the cell wall in gram-negative bacteria. The periplasmic space contains a few specific proteins; apart from those it is effectively part of the medium, as the cell wall is permeable to small molecules.

1.2.1 The antenna system and the photosynthetic unit

A reaction centre could in theory have just one chlorophyll molecule. The drawback would be that the intensity of the available light would not be enough to operate it at a reasonable speed, as the following numerical exercise shows. Note the application of Einstein's law of photochemical equivalence, which states that the primary act in a photochemical process is the absorption of one quantum by one molecule.

Figure 2.1 shows the absorption spectrum of Chl a in 80% acetone. It absorbs in a wavelength range from 400 to 700 nm. The absorption coefficients at the blue and red maxima are very high, but very sharp, so that the average absorption in the range is only approximately a third of the red peak, say 30,000 litres mol^{-1} cm^{-1}. The units of an absorption coefficient can be rearranged into the form of a cross-sectional area, thus: 3×10^7 cm^2 mol^{-1}. A single molecule has an apparent cross-section of 0.005 nm^2.

Full sunlight falling vertically amounts to approximately 1 kW m^{-2}, and nearly half of this energy is within the wavelength range to be captured by chlorophyll. The amount of energy falling per second on the apparent cross-sectional area of chlorophyll is 2×10^{-18} J. One quantum of light energy corresponding to a wavelength of, say, 550 nm has a value of 3.6×10^{-19} J. Therefore a single-chlorophyll reaction centre would only carry out five or less reactions per second. In fact the observed rate of turnover of the intermediates of the photosynthetic light reaction in green plants is 200 per second, and even in photosynthetic bacteria it is 50 per second.

Most temperate plants are, moreover, adapted to perform photosynthesis at maximum speed with light fluence rates some 1/20 that of full sunlight, and bacteria with very much less. By grouping several hundred chlorophylls together to form an antenna for each reaction centre, the turnover rate of some 200 per second is achieved. The system of specialised, cooperating chlorophylls is termed a photosynthetic unit.

Reaction centres are virtually always found with some antenna proteins around them. Such antennae are intrinsic membrane proteins. The complete antenna may consist of a large array of intrinsic membrane proteins, or a sub-

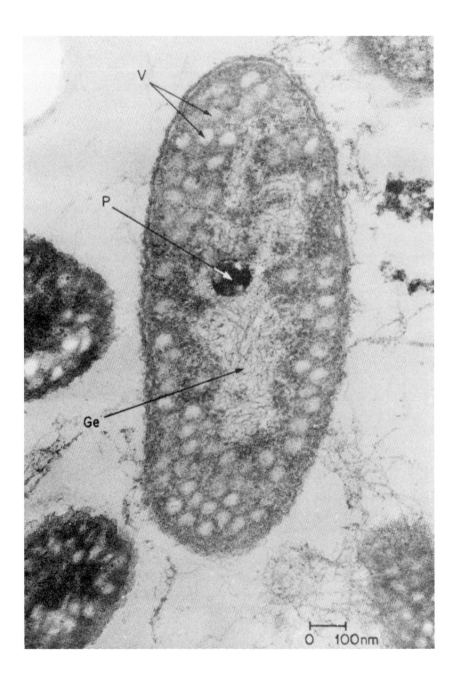

Figure 1.4.   (a) (*see legend on opposite page*)

stantial extrinsic structure attached to the membrane but projecting into the protoplasmic phase.

Survey of cell types

Purple bacteria have intrinsic antennae. The cell membrane tends to proliferate, and to form ingrowths into the cell. These may be in the form of extensive tubes, or folded sheets (Fig. 1.4a,b). When the cells are broken, the tubular type of invagination seals up, giving a preparation of chromatophores. Chromatophores are inside-out with respect to the cell membrane in that their insides are E space.

Green bacteria have extrinsic antennae known as chlorosomes on the inside (P) surface of their cell membranes (Fig. 1.5). These are large assemblies of some 10,000 BChl c molecules attached to proteins and contained in an envelope made of protein in a lipid monolayer (not a typical membrane). Chlorosomes

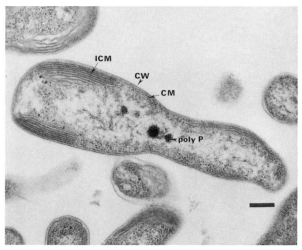

(b)

Figure 1.4. Purple bacteria. (a) Chromatophores. Section of *Rhodopseudomonas sphaeroides* (Athiorhodaceae) grown photosynthetically at low light intensity. The cell membrane proliferates into tubes or vesicles (V), some 60 nm in diameter. Also shown is polymetaphosphate deposit (P) and the region containing the DNA genome (Ge). Stained with lead hydroxide. Electron micrograph by courtesy of Dr G. Cohen-Bazire, University of California at Berkeley. (b) Lamellar structures. Thin section of *Rps palustris* cell grown at 900 lux. Large stacks of thylakoid-like membranes (intracytoplasmic membranes, ICM), the cell membrane (CM) and the cell wall (CW) are clearly resolved. Polyphosphate (poly P) granules and DNA are visible in the central region. At the right, smaller arrays of ICM are visible in the budding daughter-cell. Bar, 0.2 μm. Micrograph by courtesy of Dr Amy R. Varga, University of Illinois at Urbana-Champaign and reproduced by permission of the American Society for Microbiology Publications Department. From Varga, A.R. & Staehelin, L.A. (1983) *J. Bacteriol.* **154**: 1414–1430

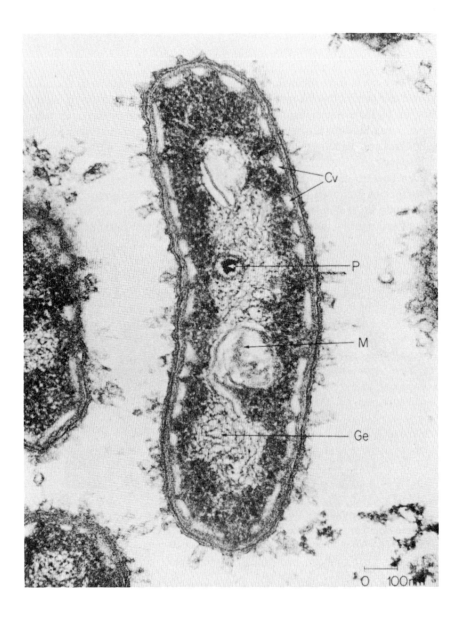

Figure 1.5.   Green bacteria. Section of *Chlorobium thiosulphatophilum* (Chlorobacteriaceae), showing chlorosomes (*Chlorobium* vesicles, Cv). M is the mesosome, an enigmatical structure that communicates with the exterior and may be involved in DNA transfer. Other details as in Fig. 1.4a

are connected to the reaction centres in the membrane. Because the BChl c is dissolved by the organic solvents used in most electron microscopic procedures, the chlorosomes used to appear as spaces known as *Chlorobium* vesicles.

Blue-green bacteria so far as can be seen do not have photosynthetically active outer cell membranes, but develop a special internal set of flattened membranes, vesicles containing E space, known as thylakoids (Fig. 1.6). These thylakoids, like the cell membranes of green bacteria, also have large particles of pigment-carrying proteins attached on their P side. However, in this case the pigments are linear tetrapyrroles (phycobilins) and the particles are known as phycobilisomes.

Eukaryotic cells contain organelles, such as the nucleus, mitochondria, endoplasmic reticulum and, in photosynthetic cells, chloroplasts (Fig. 1.7a, b). Chloroplasts are like cyanobacterial cells in many ways. They have a double (non-photosynthetic) envelope (the space between the two envelope layers is an E space) and the stroma of the chloroplast (which contains more than half the leaf protein) is a P space. Thylakoid membranes are contained within the chloroplast. Phycobilisomes are found on thylakoids of red algae, and there are loose phycobilins in the lumina of Cryptophycean algae (Chromophyta). In all other chloroplasts, including those of the higher plants, all the antennae are intrinsic membrane proteins.

In most chloroplasts there is a tendency of thylakoids to associate by appression in various ways, leading to the granal structure found in most chloroplasts of higher plants. There are grounds for believing that the thylakoids are connected so that the insides (E space) of all the thylakoids of a granal system are a continuous compartment. It is believed that the appressed thylakoids are held together by ionic forces, such as magnesium ions between negatively charged antenna complexes.

There are several protein particles found in the thylakoid membrane. These are photosystem I (PSI) and photosystem II (PSII) complexes, an antenna complex that is usually attached to PSII, an electron transport complex (cytochrome b_6f particle) and an ATPase complex. These particles are all connected functionally, by diffusible molecules and ions such as plastoquinone, plastocyanin and hydrogen ions, but it is surprising that they appear to be well separated: PSII and its antenna are in the appressed regions (the grana); PSI is in the stroma thylakoids (and on the outside thylakoids of the grana); and the cytochrome b_6f particle is located in the 'frets' between the granal 'discs' and the stroma thylakoids. The membrane and the lumina of the thylakoids in any one chloroplast are continuous, and their curiously spiral topological relationship is shown in Fig. 1.8.

A major technique in investigating the segregation of the complexes (see above) has been freeze-fracture electron microscopy. The method is to freeze a chloroplast in a block of ice, and shave it with a cold knife. Some ice is allowed to sublime from the cut surface, and this uncovers protein particles on the

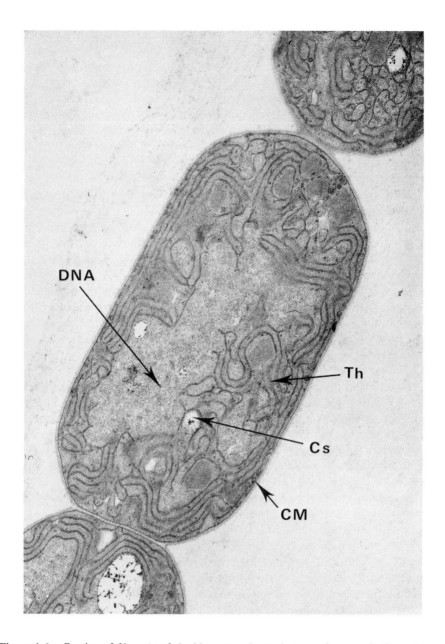

Figure 1.6. Section of filament of the blue-green bacterium *Anabaena cylindrica*. The thylakoids (Th) can be seen ramifying in the periphery of the cell bounded by the cell membrane (CM) and containing a central region with DNA. The polyhedral inclusions are probably carboxysomes (Cs), aggregates of RuBisCO. The KMnO₄ fixative has not preserved details of phycobilisomes. Electron micrograph by courtesy of Dr T.H. Giddings, Jr, Dept of Molecular, Cellular and Developmental Biology, University of Colorado at Boulder

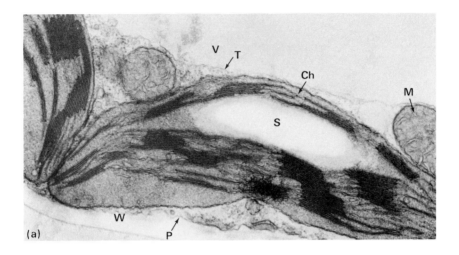

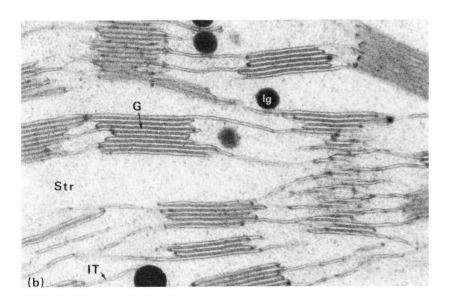

Figure 1.7.   Chloroplasts. (a) Section of part of a mesophyll cell from *Nicotiana*, showing the principal organelles. Ch, chloroplast with starch grains (S). V, vacuole. T, tonoplast membrane. W, cell wall. P, plasmalemma (outer cell membrane). M, mitochondrion. Courtesy of Dr E. Sheffield. (b) Section of a spinach chloroplast to show the construction of grana (G) and intergranal thylakoids (IT). Str, stroma. lg, lipid globule (osmiophilic globule, plastoglobulus). Courtesy of Dr L.A. Staehelin. Reproduced from Staehelin (2) by permission of Springer-Verlag

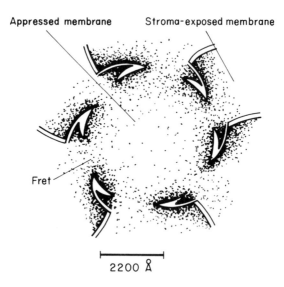

Appressed membrane   Stroma-exposed membrane

Fret

2200 Å

Figure 1.8. Construction of granum. The granum is made up of stacked thylakoid discs, each of which connects to the intergranal thylakoids with a helical twist and a 'fret' region. After J. Whitmarsh (166)

surfaces of membranes ('etching'). Other regions may show particles within the membrane: one looks for regions where the membrane has split along its central plane (between the hydrophobic tails of its lipids). Particles project from the fracture face and can be examined (as a replica of the surface) in the electron microscope. At the same time, particles projecting from the normal surfaces of the thylakoids can be made visible by freeze-etching the ice (see Fig. 1.9). Staehelin and co-workers (1,2) catalogued the particles by size, and measured their distribution in the stacked and unstacked regions (see Fig. 1.10a, b).

The segregation of particles is known as lateral inhomogeneity (3), and has two important exceptions. First, if the thylakoids are unstacked by removal of cations from the suspending medium, the distribution of particles becomes more uniform. The same effect happens in life when the PSII antenna that maintains the stacking becomes phosphorylated and migrates into stromal thylakoid regions. Secondly, there is a second population of PSII that is not confined to the grana, and which has a smaller antenna. The migrating antenna complex, when it is phosphorylated, may attach to this second population.

1.2.2 The protoplasmic phase

Attention in this chapter has been mainly focussed on the membranes of the photosynthetic apparatus. There are three non-membranous photosynthetic

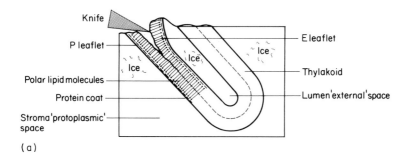

(a)

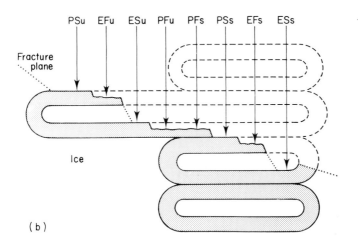

(b)

Figure 1.9.   The technique of freeze-fracture electron microscopy. (a) Knife splits ice block containing the specimen, which fractures between the two lipid leaflets of the membrane. Sublimation of the ice (etching) reveals the surfaces of unfractured membranes. A replica is made by plating on the etched and fractured surface. (b) The eight different types of membrane exposure formed in freeze fracture and etching. P and E refer to the protoplasmic- and external-phase sides of the membrane (the stromal and luminal sides in thylakoids), S and F refer to surfaces created by etching (normal surface) and fracturing, respectively, and u and s refer to membranes that are appressed or stacked (s) or unstacked (u)

structures that are worthy of note. One has been referred to above, the phycobilisome. There is another aspect to phycobilisomes in some cyanobacteria, that is, the use of the protein part of the phycobilisome as a store of nitrogenous food. The same point may be made about the enzyme RuBisCO (ribulose bisphosphate carboxylase), which is the enzyme that combines carbon dioxide

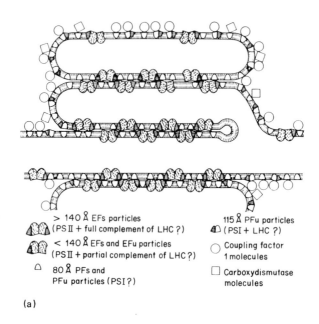

> 140 Å EFs particles
(PS II + full complement of LHC ?)

< 140 Å EFs and EFu particles
(PSII + partial complement of LHC ?)

80 Å PFs and
PFu particles (PSI ?)

115 Å PFu particles
(PSI + LHC ?)

Coupling factor
1 molecules

Carboxydismutase
molecules

(a)

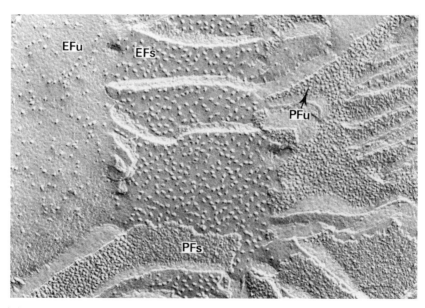

(b)

with the acceptor ribulose bisphosphate (RuBP). Cyanobacteria package this enzyme into detectable particles called 'carboxysomes', and algal chloroplasts go further and form large bodies visible in the light microscope and known as pyrenoids, consisting chiefly of this enzyme.

The third body is the starch grain, which is found well inside the chloroplasts of higher plants. In non-photosynthetic tissue starch grains are in fact chloroplasts which have never formed thylakoids. In many algae, however, starch grains are found on the outside of the chloroplast.

1.3 Prokaryote–Eukaryote, a Question of Size

Prokaryotic cells are generally small, at least in one dimension, and this may be related to the rate of diffusion of materials between enzymes in the protoplasm. Chemical processes are carried out in steps, and each step is catalysed by an enzyme. 'Enzymes are proteins which activate specific substrates and hence catalyse specific reactions' (M. Dixon). Substrates have to diffuse from one enzyme to the next. The rate of an enzyme-catalysed reaction is dependent on substrate concentration up to a limit (saturation), temperature and enzyme concentration, and also on activators and inhibitors (effectors) of different classes. In many cases these effectors are produced by the same or other enzymes, and this provides for regulatory systems, in which the rate of an overall process is regulated to meet the needs of the organism at the time. The word pathway can be used loosely to indicate the sequence of steps in which an important product is formed from a given starting point, but it can be usefully restricted to a section of the above sequence of steps in between particular reactions where the rate is regulated. Such a system frequently involves energy expenditure, and amounts to a servo-system in engineering terms. Servo-systems, because they depend on feedback, are liable to oscillate or become unstable, and the small size of primitive cells is one way to keep the feedback under control.

When we consider eukaryotic cells, an immediately obvious feature is their much greater size. This size is achieved by keeping certain processes in small membrane-bound organelles, which in many ways resemble symbiotic prokarytic cells. These are the energy-transducing organelles, mitochondrion and chloroplast, and also the site of the principal genetic store, the nucleus.

Figure 1.10. Visualisation of intrinsic membrane protein particles by freeze fracture and etching. (a) Diagram showing the populations of particles found. (b) Example of a freeze-fracture micrograph of spinach chloroplast thylakoids. The flat, partly circular membranes of two grana stacks (left and right) are shown connected by a stroma thylakoid. The typical appearance of the complementary pairs of faces PFs and EFs (see Fig. 1.9b) and PFu and EFu are shown. Note the large (16 nm) aggregate particles on the EFs face. Reproduced by courtesy of Dr L.A. Staehelin. From Staehelin, L.A., Armond, P.A. & Miller, K.R. (1976) *Brookhaven Symp. Biol.* **28**: 278–315

The large volume of the leaf cell is largely illusory; the actual cytoplasm (protoplasm that is not part of an organelle) is a thin layer inside the cell wall, and most of the bulk is the watery vacuole. The protoplasm of the chloroplast is known as the stroma. There is of course no distinction between stroma and cytoplasm in prokaryotes, and the division in eukaryotes (made by the double chloroplast envelope) allows the processes involving consumption of ATP (growth, movement, cell division) to be separated from photosynthesis. It also allows the cytoplasm to be specialised for importing and exporting material from and to other cells in multicellular plants. This is particularly important for photosynthesis in those plants known as C4, in which there is cooperation to move material between adjacent cells effectively, so as to concentrate carbon dioxide in the chloroplasts. It also provides for the differentiation of the higher plant into its different organs with different metabolic requirements, connected by the vascular system conducting water in xylem tissue, and metabolites, particularly sucrose, in the phloem.

Further Reading

Excellent micrographs and discussion of thylakoid and other structures in chloroplasts will be found in:

Manton, I. (1966) Some possibly significant structural relations between chloroplasts and other cell components. In Goodwin, T.W., ed., *Biochemistry of Chloroplasts*, vol. 1, Academic Press, London. pp. 23–47.
Hoober, K. (1984) *Chloroplasts*. Plenum, New York, chap. 2, pp. 19–45.

Chapter 2

Pigments

A pigment is a molecule that absorbs light and therefore appears coloured. The absorption of light is the necessary first step in any photobiological process (the Grotthuss–Draper principle). There is a variety of pigments found in photosynthetic systems, and further variety is imposed on them by their association with many different proteins. It will be apparent, in fact, that what may begin as a study of pigment finishes as a study of protein; not for nothing was the name 'protein' derived from 'pre-eminent'. The chapter begins with an outline of the pigments, then describes the pigment–protein complexes and their associations in photosynthetic systems. The photophysical and photochemical behaviour follows, and leads to an account of the operation and cooperation of the pigment molecules in their protein locations.

2.1 Outline of the Pigments

There are two principal groups of pigment that are important in photosynthesis, the tetrapyrroles and the carotenoids. The tetrapyrroles include the chlorophylls (Chl) and bacteriochlorophylls (BChl), which are macro-cyclic, and the phycobilins which are linear. Both groups are formed via the cyclic tetrapyrrole protoporphyrin IX (see Chapter 12) as are the haem groups found in cytochromes and catalase. The chlorophylls and bacteriochlorophylls are modified by esterification of the acid groups in the precursor, and one of the esterifying alcohols is a polyisoprenoid (derived from a repeating branched C5 unit) arising from a precursor of the carotenoid pigments. This has the effect of making the pigments insoluble in water.

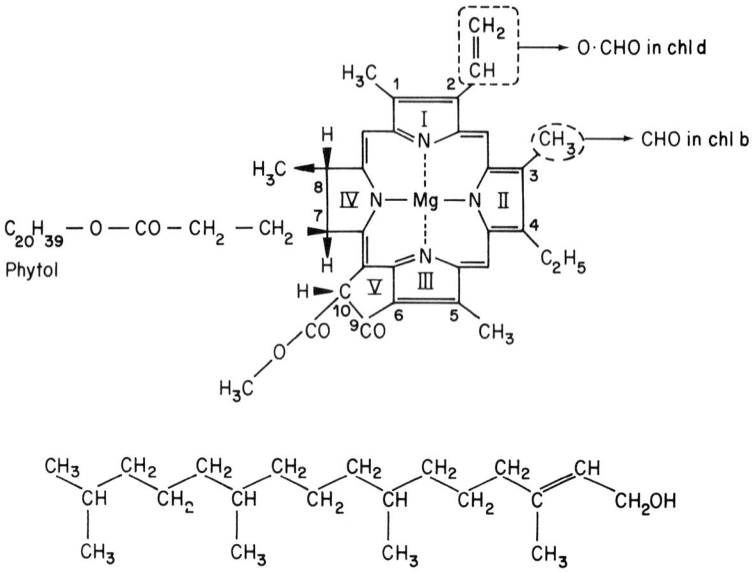

phytol

Chlorophyll a, b, d,

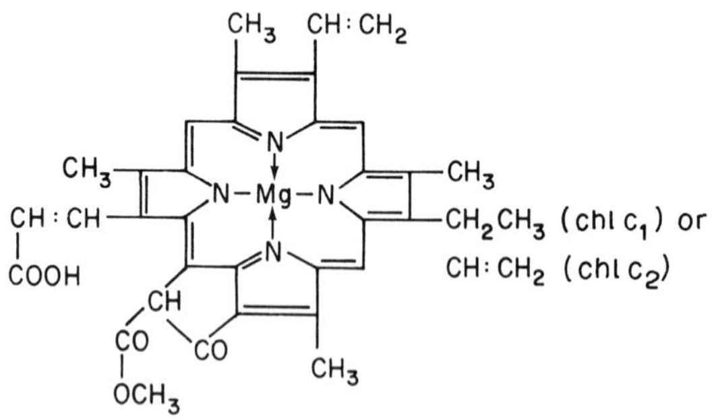

(a)

Chlorophyll c

Figure 2.1. Formulae and absorption spectra of chlorophylls. (a) Chlorophylls a, b, c and d. (b) Bacteriochlorophyll a, b, c and g. (c) Absorption spectra of Chl a, b and d, and BChl a and c. Reproduced from Smith & Benitez (4) with permission

2.1.1 Tetrapyrroles

Cyclic tetrapyrroles: the chlorophylls and bacteriochlorophylls

Chlorophylls and bacteriochlorophylls differ principally in that bacterio-chlorophylls are saturated in rings II and IV, the chlorophylls only in ring IV. They are described in Fig. 2.1. The two never occur together. Bacterio-chlorophylls are only found in the green and purple bacteria (which do not produce oxygen). The primary pigment in any species is either BChl a (or BChl b) or Chl a. Cyanobacteria (which produce oxygen) have Chl a together with phycobilins as accessory pigments (5).

Chlorophyll b is found in virtually all higher plants, although mutants can be obtained without it. There is more variation among the algae, which can be classified by means of their accessory pigment systems. The Chlorophyta (green algae, including euglenoids) contain Chl b, while the other main group,

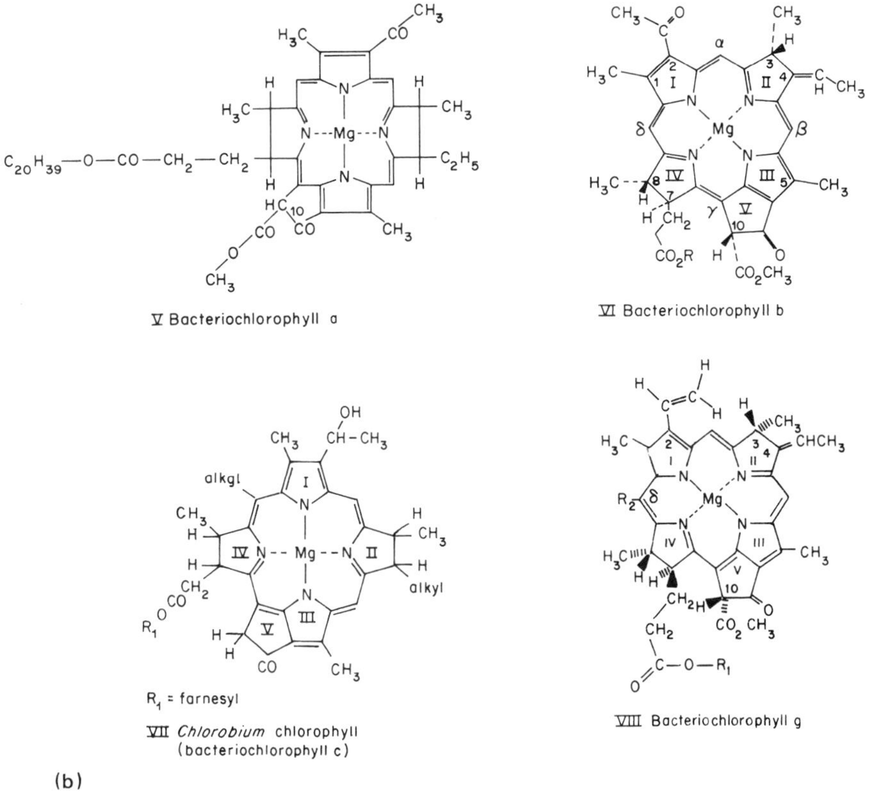

V Bacteriochlorophyll a

VI Bacteriochlorophyll b

R₁ = farnesyl

VII *Chlorobium* chlorophyll
(bacteriochlorophyll c)

VIII Bacteriochlorophyll g

(b)

Figure 2.1. (b) (*see legend on opposite page*)

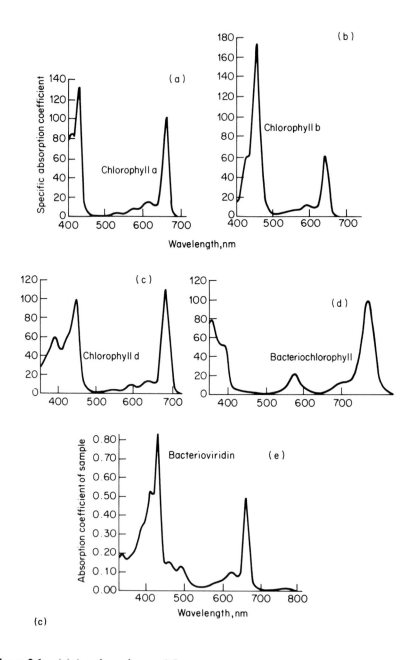

Figure 2.1.   (c) (*see legend on p. 24*)

apart from the above-mentioned Rhodophyta, the Chromophyta, possess Chl c_2 with or without Chl c_1. The Cryptophyceae (considered as a subgroup of the chromophyta) have Chl c_2 and either phycocyanin or phycoerythrin in the thylakoid lumina.

Chlorophylls may be extracted from most green plants by means of 80% aqueous acetone, although it may be necessary to remove water-soluble pigments with lower concentrations of acetone first. Chlorophylls and xanthophylls, but not all the carotenes, are extracted by this method, and can be estimated spectrophotometrically at 663 and 645 nm. Other materials need different solvents and hence different spectrophotometric routines.

Pigments can be purified by chromatography, of which high-pressure liquid chromatography (HPLC) is the best for speed, resolution, and protection against oxidation by air. Suitable media are cellulose, sucrose or polyethylene powder, but not silicic acid, which catalyses oxidations resulting in multiple chlorophyll bands.

Chlorophyll from which the magnesium is absent is termed phaeophytin (Ph), and it can be formed accidentally during operations with extracts of chlorophyll in solvents, particularly if they are acidic and exposed to light. However, there is genuine phaeophytin present in the reaction centre of PSII of green plants, and in purple bacteria (bacteriophaeophytin, BPh), where it appears to be an early acceptor of the electron in the photochemical process, forming the radical anion $Ph^{\cdot -}$.

In the green and purple bacteria the primary pigment is BChl a, except for a few species such as *Rhodopseudomonas viridis* where BChl b replaces it (but note that most other *Rps* spp. have BChl a). The long-wavelength absorption maximum of BChl a is displaced considerably by its attachment to protein, and its protein-bound forms are P860 to P890. Bacteriochlorophyll b has an even longer wavelength of maximum absorption, and the protein-bound form B1016 is the longest-wavelength absorbing pigment in photobiology. A rationale for these wavelengths is that they represent adaptations to life at different levels in a stagnant freshwater pond, where green algae absorb visible light up to 700 nm, and photosynthetic bacteria at the top of the (anaerobic) mud use the fraction of light in the range 700–1000 nm. The big differences between the absorption maxima of bacteriochlorophyll in free solution and in these proteins indicates the strong influence of the protein environment; there is much less difference in the corresponding complexes of chlorophyll.

In the green bacteria, BChl a is found in the intrinsic complexes, but the chlorosomes contain the accessory pigments BChl c, d or e, as well as some of the BChl a. Bacteriochlorophyll c used to be known as bacterioviridin, or *Chlorobium* chlorophyll 660. The newly discovered species *Heliobacterium chlorum*, which cannot be easily classified with any other group of photosynthetic bacteria, has BChl g.

The formula, obtained in an impressive study (240) at the Argonne National Laboratory, Illinois, and shown in Fig. 2.1b, shows an ethylidene group at position 4. The authors argued that a pigment of this type can give rise either to chlorophylls, in which ring II (B, new style) returns to an unsaturated state, or to bacteriochlorophylls, in which both rings II and IV are saturated. They point out that in several cases, such as the haem of C-cytochromes, and the phycobilins, there are covalent bonds between the protein (usually by means of cysteine groups) and the groups related to the vinyls on positions 2 and 4 of protoporphyrin IX. The variety of chlorophylls may derive from the different fates of an essentially common precursor like BChl g bound to a protein.

In all cases of cyclic tetrapyrrole pigments (as opposed to the linear phycobilins), there is no covalent bonding with the proteins. It is probable that in the majority of cases the magnesium atom coordinates with a histidine nitrogen atom in the protein.

Linear tetrapyrroles (bilins)

Porphyrins can be degraded by the oxidation and removal of methene bridge δ (eliminated as CO); a linear tetrapyrrole results. This is the means by which mammalian liver degrades haemoglobin and excretes the pigments bilirubin, biliverdin and others in the bile, and therefore linear tetrapyrroles are known as bile-pigments or bilins even when they are of plant origin.

In photosynthesis, bilins complexed with proteins (phycobiliproteins) form phycobilisomes, attached to the P side of thylakoids in the blue-green bacteria (Cyanobacteriaceae or Cyanophyta) and red algae (Rhodophyta). They do not occur in higher plants. Phycobilins are grouped into the phycoerythrins, which absorb maximally at about 565 nm, appearing red, and phycocyanins and allophycocyanins, which absorb at about 620 and 650 nm respectively, appearing blue. The prefix C- or R- indicates the source; thus C-phycoerythrin comes from cyanobacteria. Allophycocyanin has been found in isolated (i.e. phycobilisome-free) PS II preparations from blue-green bacteria, where it may function as a link between the phycobilisome and the intrinsic antenna of PSII. It is not a photochemically active component.

Phycobiliproteins are made up of subunits (see Section 2.2.1) which contain chromophore prosthetic groups, usually one or other of phycocyanobilin (PCB) or phycoerythrobilin (PEB); there is a minor variety, phycourobilin (PUB), that occurs with PEB in R-phycoerythrin. These bilins are attached to the protein subunits by thioether bonds to cysteine residues, and by hydrogen bonds from NH groups to aspartate residues (Fig. 2.2) as shown by X-ray crystallography on a cyanobacterial phycocyanin (241).

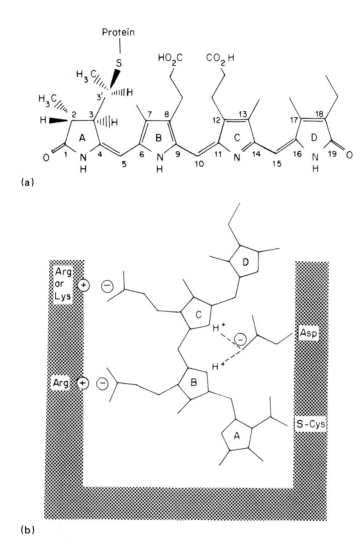

(a)

(b)

Figure 2.2. (a) Chemical structure of the phycocyanobilin chromophore, after Glazer, A.N. (1984) *Biochim. Biophys. Acta* **768**: 29–51. (b) Diagram showing the extended conformation of the phycocyanobilin chromophore and its interactions with the protein. Taken from the X-ray crystallographic solution of Schirmer et al (241); the stereoscopic computer-generated diagrams in the original show that rings A–C are virtually superposable, but ring D is twisted by a varying amount in the three locations in the phycocyanin of *Mastigocladus laminosus*. The form of the protein is shown conventionally; the original illustrations show the interactions of the different polypeptide chains

2.1.2 Carotenoids

Carotenoids are formed by the isoprenoid path described in Chapter 12. Carotenes relevant to chlorophyll-dependent photosynthesis are based on a C-40 structure. The carotenes can be oxygenated in various ways to give a great diversity of xanthophylls. To some extent this is reversible and may be part of the operation of their protective role against oxygen (see below). Because carotenoids absorb in the chlorophyll 'window' they have a great influence on the perceived colour of the organism, as in brown algae, purple bacteria etc. To some extent they are able to pass excitation energy to chlorophyll and hence act as accessory pigments. A more general role is protective. The excited state of chlorophyll has a normally short lifetime and is relatively immune from attack by agents such as oxygen; however, the normal 'singlet' excitation state can change to a 'triplet' state with a longer lifetime. Carotenoids appear to deactivate the triplet state of chlorophyll and either annihilate the excitation (as heat) or combine with oxygen sacrificially, as in the xanthophyll cycle (see below).

Carotenes (hydrocarbons) (see formulae in Fig. 2.3) are present in most photosynthetic organisms; in algae and higher plants β-carotene is predominant, attached to the reaction centres of PSI and PSII. Alpha-carotene and ϵ-carotene are minority forms. Both ends of the chain are cyclised. Gamma-carotene (one end cyclised) is widely found in green bacteria, while in the purple bacteria carotenes as such have not been established, and their carotenoids are open-chain, methoxylated structures.

Oxygenated carotenes (xanthophylls) (see formulae in Fig. 2.4) are virtually universal. In higher plants and green algae the principal form is lutein, with some violaxanthin and neoxanthin. These last two have epoxy groups, and take part in the xanthophyll cycle (Fig. 2.5). This is associated with PSII in green

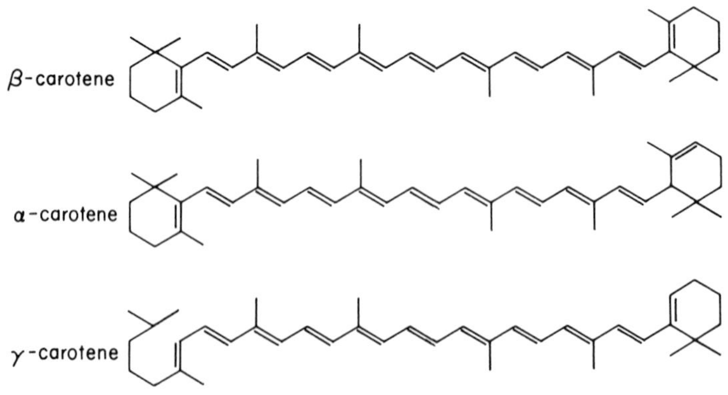

Figure 2.3. Formulae of carotenes. Singly cyclised carotenes are characteristic of the Chlorobacteriaceae

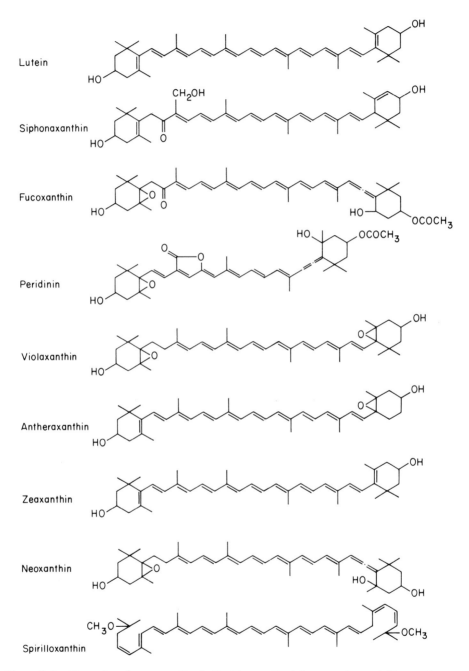

Figure 2.4. Formulae of some xanthophylls. Non-cyclised forms such as spirilloxanthin are characteristic of the purple bacteria

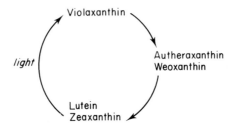

Figure 2.5. Xanthophyll cycle. The slow epoxidation and its reversal is not known to have any function in photosynthesis

plants, and may be a protective system against reactive oxygen. In the light there is conversion of lutein and zeaxanthin to violaxanthin by an epoxidation reaction. In other tissues epoxides are commonly formed by attack of active oxygen or superoxide on unsaturated lipids (for example during the formation of prostaglandins). Since so much unsaturated lipid exists in the thylakoid, a protective system is called for. Violaxanthin is returned via antheraxanthin or neoxanthin to the original forms, by means of a reducing system requiring stroma activity. The cycle takes some three minutes to complete, and therefore is secondary to the main process of photosynthesis.

In the other groups of algae common forms of xanthophyll associated with accessory pigment complexes are siphonoxanthin (some groups of the Chlorophyta), fucoxanthin (with Chl c_1 and c_2 in Phaeophyta and diatoms) and peridinin (with Chl c_2 in dinoflagellates): see the review by Anderson and Barrett (5).

In the bacteria, carotenoids are found in reaction-centre complexes of purple non-sulphur bacteria (but not in those of the related green bacterium *Chloroflexus*) and in the antenna complexes. They have rarely been identified, and are in any case more variable than in eukaryotes. Spirilloxanthin is a known component of an antenna complex in *R. rubrum*. Both in oxygenic and anoxygenic organisms, more xanthophylls are present than can be accounted for by the complement of known complexes. Often the analysis of carotenoids yields proportions less than one molecule per particle, suggesting that (i) carotenoids are only loosely attached and are lost when the complexes are isolated, (ii) carotenoids are freely soluble in the lipid membrane and attach randomly to protein complexes or (iii) carotenoids are not specific in their binding to protein complexes. These possibilities are not mutually exclusive, and it may be noted that X-ray crystallographic analyses of reaction centres do not show carotenoids, suggesting that they do not have the precise location in the molecule that is so striking in the case of bacteriochlorophyll.

Carotenoids are responsible for certain optical effects, notably the 515/518 electrochromic shift that allows the measurement of the electric field across the chromatophore or thylakoid membrane in vivo, and the C550 fast transient in PSII.

The esoteric photosynthesis of *Halobacterium* is based on bacteriorhodopsin, which is formed by the pigment retinal binding to the protein bacteriopsin. Retinal is a C-20 aldehyde derived from carotene, and its operation is discussed in Chapter 4. There is no accessory pigment; each retinal group acts independently.

2.2 Pigment–Protein Complexes

Chlorophylls and phycobilins are not believed to occur loose, dissolved in the lipid membrane, but rather as conjugates with polypeptides. The rationale for protein attachment is that efficient energy transfer from an excited to an adjacent unexcited molecule requires that the distance between them and their orientation angles should be relatively precise, in order to minimise loss of energy by fluorescence or radiationless de-excitation. A second reason is that attachment to protein provides each pigment molecule with a specific environment which alters its wavelength of maximum absorption, thus providing the means for a graded series of pigments covering a large part of the spectrum of the incident light, and conducting energy towards the photoactive pigment molecule at the active centre.

The pigmented polypeptides are usually found in complexes, that is to say in particles made up of a number of the same or different polypeptides. In this section we will examine the following levels of organisation.

(a) Large aggregates of pigmented proteins acting as accessory complexes (also called antenna or light-harvesting assemblies) extrinsic to the membrane as in the cases of the phycobilisomes of the cyanobacteria and the chlorosomes of the green bacteria.

(b) Intrinsic-protein assemblies (whose size can probably be varied in response to conditions) made up of antenna complexes surrounding embedded reaction-centre complexes. This describes the assemblies in purple bacteria and PSI and PSII of green plants.

(c) Reaction-centre complexes (core complexes) and light-harvesting complexes (LHC); these are obtained as subdivisions of (b), by means of detergents, which dissolve the lipid of the membrane and stabilise the hydrophobic protein. Complexes are resolved by such biochemical separation procedures as centrifugation or electrophoresis.

The logical distinction between antenna and reaction-centre complexes in purple bacteria is easy, because the reaction centre only carries the pigments necessary for its operation. As will be described, the photosystems of green plants, after removal of antenna (LHCI and LHCII), still contain many chlorophylls (approximately 60 and 100 respectively), so that, assuming up to six 'photoactive' chlorophyll molecules, there are appreciable 'internal antennae' in the core complexes.

The chief virtue of the nomenclature whereby PSI, for example, is resolvable into the antenna LHCI and the core complex CCI, is that it identifies biochemically the particles seen in freeze-fracture micrographs of the thylakoid (see Chapter 1). It is also the case that CCI and CCII are much less variable than the antenna complexes, throughout the range of green plants.

2.2.1 Extrinsic antenna complexes

Chlorosomes

These extra-membrane particles contain 10,000 BChl c molecules as well as some γ-carotene and some BChl a. Bacteriochlorophyll c when extracted absorbs at 660 nm, but in the chlorosome, bound to protein, it has a maximum at 740 nm. Using the well known notation Bλ to indicate a BChl complex that is not photoactive, the chlorosome contains B740. There is a single type of polypeptide (see Fig. 2.6), which binds seven BChl c molecules.

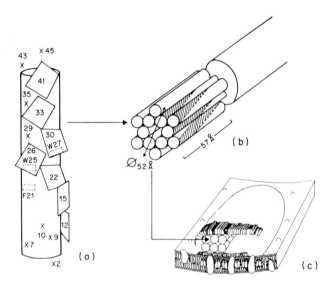

Figure 2.6. Structure of chlorosome of green bacteria, showing (a) seven BChl c molecules per rod element, (b) assembly of rod elements into bundles, and (c) assembly of bundles into chlorosome. Reproduced by permission of Professor H. Zuber and Elsevier Publications Cambridge. From Zuber, H. (1986) *Trends Biochem. Sci.* **11:** 414–419

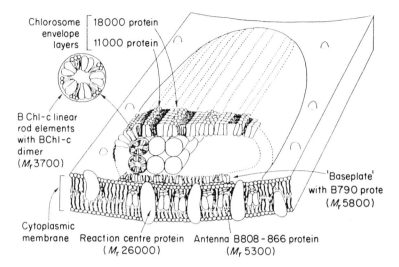

Chlorosome ⌈ 18000 protein
envelope
layers ⌊ 11000 protein

B Chl-c linear
rod elements
with BChl-c
dimer
(M_r3700)

'Baseplate'
with B790 prote
(M_r5800)

Cytoplasmic
membrane   Reaction centre protein   Antenna B808 - 866 protein
(M_r26000)   (M_r5300)

Figure 2.7. Structure of chlorosome. The relationship of the chlorosome bundles to the membrane and other Bchl–protein complexes. Reproduced by permission of Professor R.C. Fuller and the American Chemical Society. Copyright (1984) American Chemical Society. Reprinted from Feick, R.G. & Fuller R.C. (1984) *Biochemistry* **23**: 3693–3700

Twelve of these protein units are organised into a rod in such a way that the 84 BChl molecules trace a helical pattern (Fig. 2.7). There are some 120 such rods, and the base of this multi-fibrillar structure is connected to the membrane via a 'base-plate' complex, which is water-soluble and contains seven BChl a molecules per peptide. This base-plate complex was the first chlorophyll–protein complex to be crystallised and solved by X-ray diffraction (8). The protein is clearly globular, and holds the BChl a molecules in a basket of β-structure (Fig. 2.8).

The base-plate complex is connected to the intrinsic membrane antenna complex, B808–866.

Phycobilisomes of blue-green bacteria and red algae

Blue-green bacteria and red algae possess the photosystems I and II characteristic of higher plants, but do not possess the light-harvesting complex that accounts for up to half the complement of higher plant chlorophyll. Their antennae are provided by the extrinsic assembly of phycobilins that is attached on the cytoplasmic or stromal (P side) surface of the thylakoids, known as the phycobilisome. In most species the phycobilisomes are hemidiscoidal or hemispheroidal and contain 300–800 phycobilin molecules. The phycobilins are attached to proteins by covalent thioester bonds with cysteine. The proteins

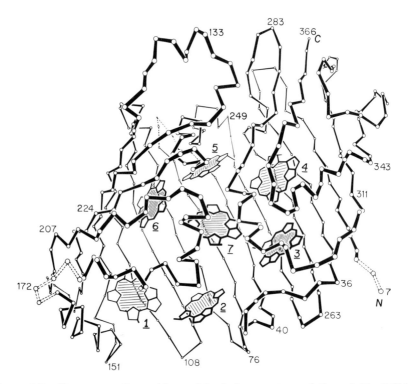

Figure 2.8. X-ray crystallographic model of the structure of the soluble BChl a–protein complex from *Prosthecochloris aestuarii*. This base-plate pigment–protein complex is water-soluble, and was the first complex to be crystallised. Although the BChl a molecules are not covalently bonded, their sharpness in the X-ray diffraction shows that their positions are precisely determined. Note the basket formed by the β-structure of the polypeptide chain. After Fenna and Matthews (9). Redrawn after Tronrud, D.E., Schmid, M.F. & Matthews, B.W. (1986) *J. Mol. Biol.* **188**: 443–554, with amino acid residues numbered according to Daurat-Larroque, S.T., Brew, K. & Fenna, R.E. (1986) *J. Biol. Chem.* **261**: 3607–3615

are globular, water-soluble heterodimers of α and β chains (16–20 kDa) that have sequence homologies. They form aggregates (trimers and hexamers) that correspond to discs seen in the electron microscope. The phycobilins are bound to the proteins at a common position (Cys-84, also Cys-155) on both α and β chains.

The discs are attached to linker polypeptides that themselves carry phycobilins. There are three principal types of disc-protein, which absorb at different wavelengths owing to their different complements of pigments: phycoerythrins or phycoerythrocyanins (absorption maxima around 560 nm), phycocyanins (620 nm) and allophycocyanin (650–680 nm). The linker polypeptide for phycoerythrin is longer (34.5 kDa) than the others and its pigments absorb at 568 nm. The linker for phycocyanin (29.5–34.5 kDa) absorbs at 620–630 nm (three types are known). The linker polypeptides govern the position of the discs so that the shorter wavelength discs are on the outside. Inside are the discs of allophycocyanin which lead to a large 90 kDa peptide and allophycocyanin B (680 nm). The complete structure worked out by Zuber (7,10) for the cyanobacterium *Mastigocladus laminosus* measures 40 × 70 nm and is drawn in Fig. 2.9.

2.2.2 Intrinsic antenna complexes

Purple bacteria

The light-harvesting complexes of purple bacteria are intrinsic membrane proteins, and have been extensively studied. Detergents have been essential in disrupting what now appears to be an extensive network of aggregated pigmented protein surrounding the reaction centres. Triton X-100 was succeeded by sodium dodecyl sulphate, and then by lauryldimethylamine-N-oxide (LDAO). The pigment allows the procedures to be followed by eye during chromatography, etc., and also indicates, by a change in the wavelength of the absorption maximum towards that of the pigment in solution (i.e. a blue-shift), any change in the tertiary or quaternary state of the protein.

Both sulphur and non-sulphur bacteria have antenna complexes that are composed of pairs of two different peptide subunits (α–β heterodimers), and carry BChl a (or BChl b in species such as *Rps viridis*). The subunits are composed of 50–60 amino acid residues each, and have a central membrane-spanning segment with hydrophilic regions at either end, which presumably project from the membrane into the P and E spaces, because these ends can be digested by proteinases. Figure 2.10 shows a model of the heterodimer in relation to the membrane. The sequences of amino acids show clear homology both between the α and β peptides and between peptides from the different species. From the protein sequence homologies it can be argued that the α and β chains originated by evolution from a common precursor, and that there is an evolutionary succession shown by the sequence homologies of the α and β subunits in the species *Rps viridis* (most primitive), *Rhodospirillum rubrum*, *Rps gelatinosa*, *Rps capsulata* and *Rps sphaeroides* (67).

Each peptide binds one or two bacteriochlorophyll molecules and not more than one carotenoid, so that the ratio of BChl a to carotenoid is either 2:1, 2:2 or 3:1. Although there are only two types of peptide chain making up the antenna in any one species, when the antenna is broken up into complexes up to three different kinds can be distinguished by their absorption maxima. It is believed that they are arranged in the intact antenna so that the longest wavelength absorbing forms are adjacent to the reaction-centre complexes, while

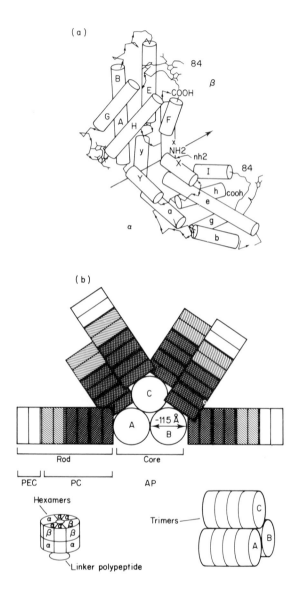

Figure 2.9.   (*see legend on opposite page*)

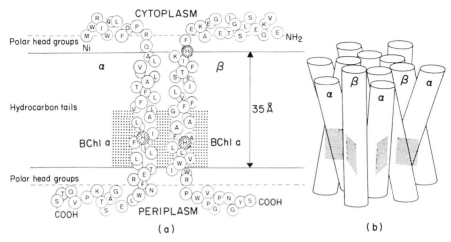

Figure 2.10. Structure of the antenna complexes of purple bacteria. (a) All the three types known have α- and β-subunits, each with three domains, the central one spanning the membrane. The α-subunit carries one bacteriochlorophyll, the β-subunit two; all three bacteriochlorophylls are complexed by interaction of the central magnesium atoms with histidine (H), shaded. The stippled area indicates an interacting pair of bacteriochlorophylls. The example is from *Rhodospirillum rubrum*. (b) The dodecameric aggregation $(\alpha_6\beta_6)$ suggested, for example, in the antenna B800–850. Reproduced by permission of Professor H. Zuber and Elsevier Publications Cambridge. From Zuber, H. (1986) *Trends Biochem. Sci.* **11**: 414–419

the shortest wavelength forms are peripheral, or furthest from the reaction centres if the assembly contains several. For example, in the Athiorhodaceae, *Rps viridis* only contains one antenna (B1015), as does *R. rubrum* (B870–890). *Rps sphaeroides* contains both B870–890 and B800–850; *Rps acidophila* contains these two plus a third (B800–820). The purple sulphur bacterium *Chromatium vinosum* also has the same three. The B870–890 antenna is common to all BChl a-containing purple bacteria (11) and corresponds to the B1015 antenna of those that contain BChl b.

Figure 2.9. Structure of the phycobilisome. (a) The α-subunit structure of C-phycocyanin from the cyanobacterium *Mastigocladus laminosus*. The α-subunit carries one bile pigment molecule, at position 84, and the β-subunit carries two, at positions 84 and 155. (Compare Fig. 2.2a.) The polypeptide chains form extensive sets of α-helical sections, labelled X, Y, A, B, E, F, G. There is extensive homology with other phycobilins in this species, and with those in red algae, for example *Cyanidium caldarum*. Reproduced from Zuber (10). (b) The phycobilisome formed by (i) formation of sets of hexamers $(\alpha_3\beta_3)$ around a linker peptide, (ii) stacking of phycocyanin (PC) and phycoerythrin (PEC) hexamer-sets to form rods and (iii) arrangement of the rods around a core of allophycocyanin (APC) trimers. Reproduced by permission of Professor H. Zuber and Elsevier Publications Cambridge. From Zuber, H. (1986) *Trends Biochem. Sci.* **11**: 414–419

The green bacteria are not easily comparable, since their antennae are the massive chlorosomes containing mainly BChl c. In *Chloroflexus aurantiacus* the intrinsic membrane antenna contains the BChl a form B808–866. In this case there are 20–25 antenna bacteriochlorophylls per reaction centre.

It has been suggested that the reason for the difference in the number of bacteriochlorophyll molecules in the antennae of the purple and green non-sulphur bacteria lies in the greater ability of the latter to grow under conditions of low light. The green sulphur bacteria are often intolerant of high light conditions, and may lack a means of safely deactivating the pigments when the excitation is excessive.

Light-harvesting complexes I and II of green plants

Green plants have a more complicated photosynthetic apparatus than either the green or the purple bacteria, in that they have two photosystems, PSI and PSII, and there are specific antenna systems associated with each.

Photosystem II is normally found associated with the antenna system LHCII. Its existence was first shown by the isolation of the Chl (a+b)–protein component, CP2, by means of detergents such as SDS followed by electrophoretic separation (12–14). It is the major antenna of chloroplasts and contains half the total chlorophyll and almost all the Chl b. Less powerful detergents allow the isolation of oligomeric forms. It now appears that there are several very similar peptides, two to four of which may occur together. These proteins are sufficiently similar to co-crystallise interchangeably (15). In addition to four molecules of Chl a and three of Chl b, there are one to two molecules of carotenoid, lutein or neoxanthin or both. The functional subunit may be trimeric, indicated both by the electron microscopy of crystals and the circular dichroism (CD) of the pigments, which shows strong trimeric interaction for Chl b. (Zuber's comparative studies on other antenna systems led him to propose a general dimeric structure for the functional subunits of antenna complexes, and he established that the phycobilisome subunit is a hexamer (trimer of dimers)(7).) A hexameric structure is not ruled out for LHCII. The protein chains of the LHCII peptides are likely to contain three membrane-spanning sections of α-helical structure. This conclusion is based on the hydropathy count, a procedure provided by Kyte and Doolittle (16) in which each amino acid residue is allocated a hydropathy index based on its partition coefficient, and these are summed in a frame which travels along the sequence; a high hydropathy extending over a sequence of 20 amino acids is likely to be a membrane-spanning α-helix. The amino acid sequence of the protein contains runs of amino acids that are known to be helix formers in other proteins.

There are at least six to seven chlorophyll molecules per peptide of LHCII, more than in the bacterial antenna, and too many for each to be bound to a specific histidine residue.

LHCIIβ has been shown to take part in an important regulatory mechanism. It is reversibly phosphorylated (on threonine residues) by a 64 kDa kinase located in the thylakoid. The activity of the kinase is controlled by the degree of reduction of plastoquinone so that the more reduced the quinone (owing to a greater activity of PSII with respect to PSI) the greater the rate of phosphorylation. There is a phosphatase that dephosphorylates LHCII at a constant rate, so on balance the degree of phosphorylation of LHCII depends on the excess photochemical activity of PSII. The effect of the phosphorylation is to diminish the activity of PSII, and it has been suggested that the increased negative charge causes the antenna to detach from the PSII reaction centre because of electrostatic repulsion. Magnesium ions inhibit the changes, probably by providing a screen and reducing the electrostatic force. The detached LCHII possibly associates with PSIIβ or even PSI, and thus regulates the distribution of energy. These findings, by Bennett and co-workers (17) have been argued (18) to be the basis of 'state I—state II' transitions, in which the relative cross-sectional area for light capture by one photosystem is increased by exposure to light preferentially absorbed by the other. Thus state I is produced by light of 700 nm wavelength. It is established that LHCIIβ becomes phosphorylated and migrates from the grana (strictly, the stacked or appressed areas of the thylakoids) to the stroma, and that about 20–25% of the LHCII particles are involved. The migration is slowed down by increasing the viscosity of the membrane by means of cholesteryl ester (19). The fluorescence at 77 K of stroma lamellae isolated from state II chloroplasts showed both an increase in F735 (fluorescence from a complex close to P700, see below) and an increase in the Chl b:a ratio. The importance of phosphorylation in causing the rearrangement of the antenna complexes is clear, as is the neat feedback mechanism that adjusts it. However, there is a problem in that LHCII in transit should be disconnected from any reaction centre, and therefore a relatively enormous fluorescence at 677 nm ought to be seen transiently while the transfer is taking place; it is not seen. To some extent appearances could be saved by supposing that LHCII does not detach from PSII reaction centres in the grana, but segregates so that the PSII centres are isolated into smaller groups (the lake is divided into smaller puddles).

Protein phosphorylation has also been shown to regulate the size of cooperating units in the purple bacteria (and in cyanobacteria, although there is no appression of membranes and therefore, probably, no lateral inhomogeneity). Allen et al suggested (20) that a 9 kDa peptide component of CCII in PSII of both chloroplasts and cyanobacteria is the key. The 9 kDa peptide has some homology with, and may be genetically related to, the N-terminal part of the (larger) LHCIIβ peptide. The phosphorylation of this peptide has been known for as long as that of LHCII (21). So far as is known the 9 kDa peptide does not carry chlorophyll although it appears to retain one histidine homologous to those in LHCII. The 9 kDa peptide is coded on the chloroplast genome (*psb*H) while LHCII is nuclear coded by multiple genes.

The complex CP29 (first reported by Camm and Green (22)) has a Chl a : b ratio of 3 : 1–4 : 1. It has been regarded as part of the antenna of PSII, and labelled LHCIIα. CP29 is now regarded as part of CCII, rather than LHCII, and fits in with the concept of energy transfer along a sequence with the longest-absorbing forms of chlorophyll nearest the PSII reaction centre. The wavelength maxima (of Chl a components) in the various complexes are in the order of their sizes: CP23 at 668 nm, LHCIIβ at 672–675 nm and CP29 at 677 nm. CP29 is comparable in function, perhaps, with linker peptides in phycobilin-containing thylakoids (and it is of the same size).

Of possible relevance to the question of the interrelationships of chlorophyll-carrying polypeptides of about 30 kDa is the Chl a/b antenna of the cyanobacterium *Prochloron didemni*. This species is an endosymbiont of ascidians, and cannot be cultured freely. A second prochlorophyte species has been recently discovered that is free-living (23). These two species, unlike all other cyanobacteria, contain Chl b and not phycobilins. There is, however, no obvious sequence homology between the Chl a/b protein and LHCIIβ of chloroplasts.

Antennae which may prove to be related to the LHC group of higher-plant chloroplasts have been isolated from various groups of algae, and have prominent carotenoid components. (CCI and probably CCII are little different from the higher-plant forms.) The dinoflagellate *Gonyaulax polyhedra* has three. The largest (70 kDa) protein carries (in order of decreasing abundance) xanthophylls (mainly diadinoxanthin and peridinin), Chl a and Chl c_2. Smaller forms which appear to form mixed aggregates have Chl a and Chl c_2, or peridinin and Chl a. The peridinin–chlorophyll proteins are water soluble and may be peripheral to the Chl c_2 proteins. Similar patterns are found in diatoms and brown algae; in the latter fucoxanthin or violaxanthin, and Chl c_1 or Chl c_2, are also found attached with Chl a to protein. Judging by the associations between the complex and the isolated PSI and PSII systems, it appears that the fucoxanthin Chl a/c_2 protein complex is an antenna for PSII, and the violaxanthin Chl a/c_1/c_2 protein complex for PSI (see the review of Anderson and Barrett (5)).

When these units form macromolecular complexes there are large spectral shifts in the pigments, suggesting powerful interactions and changes of conformation. In each case it has been shown that the protein does indeed function as an antenna, and that light absorbed by the carotenoid and Chl c is available for photosynthesis.

PSI antennae: LHCI

Although it has been known for a considerable time that the reaction centre of PSI, a pair of Chl a molecules known as P700, can be obtained in a protein complex with a variable number of other Chl a molecules, the resolution of PSI into core and antenna complexes has only been achieved since 1980. Progress followed Mullet and co-workers' isolation of an active PSI preparation from chloroplasts (24), which they regarded as a core complex, predicting the existence of a detachable antenna complex.

The antenna complex was isolated in the same laboratory by Haworth and colleagues (25) and shown to contain Chl a and b (a : b ratio 3 : 7). There were four polypeptides in the mass range 19–24 kDa. An important property was the fluorescence emission maximum at 730 nm (see Section 2.4.2). The complex was labelled LHCI and the previously known LHC was re-named LHCII.

Two PSI antenna complexes were resolved by Lam et al (26). The 730 nm fluorescence was retained by one complex, LHCIb, while the other (LHCIa) fluoresced at 680 nm and was thought to be denatured. A recent preparation by Bassi and Simpson (27) using dodecyl maltoside as the detergent has obtained LHCIa fluorescing at 690 nm (again a significant value). The CD spectrum of the LHCIa indicates that the protein is (relatively) in the native state.

These PSI antennae differ from the PSII antenna LHCII in that they are not readily detached, and do not have the association–dissociation behaviour found with LHCII and PSII. For some purposes they may be regarded as part of the PSI complex, just as the antennae CP43 and CP47 are part of the PSII complex. Nevertheless, it will be shown later in this chapter that analysis of fluorescence provides an insight into the structure of the PSI particle and the groups of chlorophyll molecules within it, such that the concept of distinct core and antenna complexes is useful (Fig 2.11).

The LHCIa and b antennae were at first judged to be immunologically non-cross-reactive with respect to LHCII (26). Later, cross-reactions were claimed (28), but recent studies employing monoclonal antibodies indicate that the only cross-reactions are between LHCIIα (CP29) and LHCI (not LHCIIβ). LHCIIα and LHCI have at least two common epitopes (242).

Algae such as *Chlamydomonas* have been shown to contain an antenna system for PSI, known as CPO. In a recent study (29), CPO has been shown to contain five polypeptides in the mass range 19–27 kDa, of which at least two (20 and 26 kDa) carry CHl a and b. Like LHCII, the 20 kDa peptide has been shown to be coded in the nuclear DNA. CPO may well prove to be homologous with LHC from higher plants.

The following summary of general principles concerning antenna complexes has been made by Zuber (7):

(a) Antenna complexes in the whole antenna are regularly arranged by specific interactions of polypeptides.
(b) The pigment molecules in the antenna complexes and antenna systems are highly ordered and form pigment clusters, for structural reasons.
(c) Pigment clusters are important for heterogeneous, directed energy transfer

between antenna complexes that are spatially separated and have different absorption maxima, to minimise the length of the random walk of energy to the reaction centre.

(d) Pigment clusters are also found within the antenna complexes in the form of exciton-coupled pairs or oligomers, constituting energy traps and providing for energy transfer within the complexes.

(e) The pigments are bound specifically to polypeptides that are generally small. The pigment organisation (see (d)) depends on the specific organisation of the polypeptides.

(f) In the antenna complexes of bacteria, including cyanobacteria, and red algae, with cyclic symmetry, there are pairs (heterodimers) of α and β subunits that are arranged as repeating subunits.

2.2.3 Reaction-centre complexes

Bacteria

Isolation of functioning reaction centres was achieved for the first time by Reed and Clayton (30) from *Rps sphaeroides* R-26, and by Gingras and Jolchine (31) from *R. rubrum* G-9, using carotenoidless mutants in each case. The *Rps viridis* reaction centre followed in 1969 (32). Detergents were used to extract the complexes from the membrane, as described for the antenna complexes. Unlike green-plant reaction centres, or those from green sulphur bacteria, purple and green non-sulphur bacterial centres have no antenna bacteriochlorophyll, and the optical analysis of the reaction sequence has been unhindered.

Reaction centres from purple non-sulphur and sulphur bacteria contain three intrinsic membrane-spanning polypeptides, known as H (heavy), M (medium) and L (light). M and L carry between them four bacteriochlorophyll molecules (BChl b in *Rps sphaeroides*, BChl a in most species), two bacteriophaeophytin (b or a as before) and one carotenoid. Subunit H does not carry any pigment. In some species there is an extrinsic, hydrophobic *c*-type cytochrome that is attached to the E side of the complex, but in others the soluble (hydrophilic) cytochrome *c* has direct access to the P870 bacteriochlorophyll. There are obvious sequence homologies between L and M, and also (but less strikingly) between those and the subunit H, and Hearst and Sauer (33) have suggested that this shows an evolutionary sequence from an ancestor in which there was only one peptide; a process of gene duplication and differentiation could have provided first the primaeval H subunit, then later a better bacteriochlorophyll-holding framework of the two peptides L and M. (The main object of their report was the striking homology between subunit L and the Q_B-protein from green-plant PSII.)

The reaction-centre complex has been crystallised, and the impressive X-ray crystallographic model produced by Deisenhofer and co-workers (34) for this type of reaction centre is discussed in Chapter 4.

The green non-sulphur species *Chloroflexus* has a reaction centre very similar to that of the purple bacteria except that there are three each of BChl a and BPh a, instead of four and two respectively, and there appears to be no H subunit or carotenoid. It would be unsound to say that *Chloroflexus* represents the primitive state of affairs in every respect, but the absence of the two components allows one to conclude that they have no essential photochemical function.

The green sulphur bacteria appear to be completely different from the above-mentioned types. A reaction-centre complex has been prepared from the green sulphur bacterium *Prosthecochloris aestuarii*, which had 35 BChl a molecules per active centre (P840). A preparation from *Chlorobium limicola* by Hurt and Hauska (35) contained one bacteriochlorophyll polypeptide of 65 kDa, strongly suggesting similarity with PSI of green plants (see below). However, it also contained cytochrome c-553 and a possible Rieske iron–sulphur protein, which (if not due to contamination) are more like the bc electron transport complex. Could the green sulphur bacterial reaction centre be an evolutionary precursor of both PSI and the bc complexes?

Reaction centres of green plants

Photosystem II of green plants has only recently been extracted as a functioning complex (see Chapter 5); it is unfortunately destroyed by electrophoresis in the presence of the detergent SDS, which has been so useful in resolving other chlorophyll–protein complexes. Milder conditions preserve two intrinsic polypeptides containing Chl a, named CP43 and CP47 from their apparent molecular masses in kDa. Inevitably, different laboratories will make different assessments of the molecular masses from their gel-electrophoreses, and use different symbols, but there is not usually any difficulty in cross-relating them. These two polypeptides carry at least 20 Chl a molecules each. Green and Camm (36) provided evidence that the photoactivity of PSII was associated with CP47, and for a time it was considered likely that CP47 carried the active centre.

Two other peptides, known as D1 and D2 (33 and 32 kDa respectively), are not known with certainty to have chlorophyll attached, but one hypothesis predicts that they should do so since they have a striking sequence homology with bacterial L and M reaction centre subunits, and therefore may carry the active-centre chlorophyll P680, in an alternative hypothesis to that of Green and Camm above (see Barber (37)). In fact a particle recently prepared (isolated) by Satoh and Nanba (38) contained D1, D2, and b-559 proteins with five Chl a and two Ph a molecules.

A fifth chlorophyll–protein complex is the antenna CP29. CP29 is part of the PSII core complex, and has no obvious relationship with the components of the

LHCII antenna (see above), although it does carry Chl b as well as Chl a. In all, there are some 200 Chl a (and some Chl b) molecules in the PSII complex.

There are two molecules of cytochrome b-559 (two subunits, 9 and 4 kDa), and a number (not agreed) of smaller peptides. Some of these may be required to carry the (unknown) electron transport mediators of the oxidising side of the PSII photoreaction. There is, however, evidence that some peptides, particularly D2, turn over rapidly by proteolysis in light, and therefore the number of polypeptides found may depend on the history of the material.

There are three extrinsic peptides attached on the E side of the complex, as demonstrated by their easy removal when inside-out thylakoid vesicles are prepared. They have apparent masses of 32, 23 and 18 kDa, and appear to protect rather than contain the binding site of four manganese atoms that are almost certainly part of the oxygen-evolving mechanism. Chloride and calcium ions are implicated in the maintenance of the structure.

A model representing the more recent concept of PSII is shown in Fig. 5.3.

Photosystem I

The prominent chlorophyll–protein complex obtained on polyacrylamide gel under denaturing conditions, CPI, represents the principal polypeptide of PSI, as shown by its absence in a mutant of *Scenedesmus* known to be lacking PSI. CPI contains two related but distinguishable peptides of around 70 kDa, and the functional core complex (CCI) is based on a heterodimer of CPI (or tetramer as has been claimed (39)). There are 50–60 Chl a and 1–2 β-carotene molecules, and the dimer carries at least one P700 active centre. P700 is probably a dimer, at least in the unexcited state, of two Chl a molecules, one of which appears to either be or to give rise to the chlorophyll derivatives Chl RC-I and Chl a′ (see Chapter 6). A more intact PSI core complex can be isolated, containing the above CPI plus two subunits believed to have a structural role, and three or more small peptides that are believed to carry secondary electron acceptors. A model for the complex and its associated antenna, LHCI, appears in Fig. 2.11.

2.3 Light Absorption and the Subsequent Fate of its Energy

2.3.1 Light absorption

The absorption of light by a pigment takes place in something of the order of a femtosecond (10^{-15} s) and involves a change in one electron in the absorbing molecule from a lower energy to a higher energy state. The energy states for a given electron are quantised, and absorption only occurs if energy is available in the light beam that corresponds with the energy required for a particular promotion. A light beam is a stream of energy packets termed quanta, the

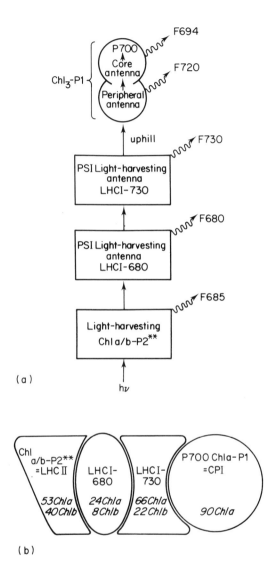

Figure 2.11. Antennae and core complexes of green-plant PSI. (a) Resolution of a preparation from barley into LHCII, two distinct LHCI fractions and a core complex. The core complex contains the P700 group and a core antenna, and a peripheral antenna. The characteristic absorption and fluorescence emission maxima are shown. (b) Model of PSI in which the complements of chlorophyll are shown for each complex. Note that LHCII would not necessarily be attached (so much) in vivo. Reproduced from Bassi & Simpson (27) by permission of Springer-Verlag

energy of each quantum depending on the frequency of the light ($h\nu$, where h is Planck's constant and ν is the frequency) or on the wavelength (hc/λ, where λ is the wavelength and c the medium-dependent speed of light). The promotion of a particular electron requires the absorption of a quantum having the required energy. In a population of molecules there is a spread of energies due to vibrations, intermolecular interactions etc., so that the absorption spectrum shows the familiar bell-shaped curve for each electron promotion.

A quantum of light of wavelength 1240 nm is 1 electron-volt of energy; an einstein (Avogadro's number of quanta, a mole-quantum) is 96.649 kJ. This is close to the lower energy limit for electronic transitions, and the absorption of (infrared) light of longer wavelengths tends to produce molecular rotations and vibrations and hence heat. The longest wavelength-absorbing pigment in biology appears to be the BChl b in the antenna complex B1016 in *Rps viridis*, and this could be taken as the limit of 'visible' light. Photosynthesis and indeed all photobiology is possible because most of the essential materials of the cell do not absorb in the visible range, and those that do, electron transport carriers such as cytochromes, are not in sufficient concentration to absorb a significant part of the incident light. Photobiological pigments can therefore absorb energy without interference over a range from below 400 nm up to the limit . This range is denoted by the acronym PAR: photosynthetically active radiation. Below 300 nm (PUVB: photobiological ultraviolet range B) absorption becomes virtually general. Not only is there a lower proportion of light for possible photosynthesis, but there is also the hazard of uncontrolled photochemistry occurring in the proteins and nucleic acids of all the cells of the organism.

2.3.2 Fate of excited pigments

A molecule that has become excited by the absorption of one quantum can proceed in several ways: fluorescence, energy transfer, non-radiative deactivation and photochemistry.

Fluorescence

Fluorescence is the re-emission of a quantum of light by the excited molecule. This is the only possible fate for an isolated single pigment molecule (of vapour), and the excited state would last (statistically) for a characteristic time, which for Chl a has been calculated to be 15 ns. The decay is of course continuous, and exponential; the lifetime is simply the reciprocal of the velocity constant k in the first-order kinetic equation for the decay of A^*

$$\mathrm{d}A^*/\mathrm{d}t = kA^*$$

Fluorescence differs from scattered light in that (i) light is scattered in the same time period as absorption takes place, (ii) no energy is lost to vibration, etc., and the wavelength of the scattered light is the same as the incident beam (neglecting Raman effects), and (iii) the intensity of scattered light as a function of wavelength is related to the refractive index of the pigment, that is, its dispersion (including the effect on the refractive index, anomalous dispersion, in the region of absorption). The two effects are similar in that for fluorescence from a single molecule, the polarisation of the emitted light is preserved, because there has been insufficient time for the molecule to move or rotate by a significant amount.

The wavelength of maximum fluorescent emission is usually longer, that is, of lower energy, than the wavelength of maximum excitation (the Stokes' shift, often about 8 nm for chlorophylls). Figure 2.12 provides a generalised expla-

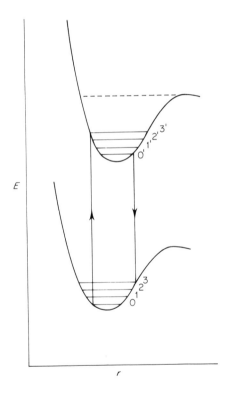

Figure 2.12. Diagram of vibrational energy levels in a chemical bond in the ground and excited states. E is the energy and r is the distance between the nuclei. The model indicates a transition due to absorption of a photon from the 0th level of the ground state to the 3rd level of the excited state: this is a 0–3' transition. Also shown is a 0–3' transition emitting a photon of lower energy. Transitions are most likely between the extrema of vibrations, as shown, and the relative positions of the two curves determines the most likely transitions, and in particular the transition that disrupts the bond (dotted line)

nation, based on the Franck–Condon principles. These are (i) that absorption and emission are very rapid processes compared to the speed of molecular vibrations, and (ii) that the promotion takes place at the end of a vibrational movement. The solid lines of the graph indicate the energy associated with a given position of the atoms of the bond, and hence show the mean bond length and the point at which the bond breaks. However such a graph only has numerical precision for a diatomic molecule, and for a molecule such as chlorophyll it is only qualitative, and indicates principles only. There is no guarantee that in any particular case the upper curve will be to the right of the lower one.

Fluorescence has at least two major applications in photosynthesis. First, the spectral dependence and lifetime can be analysed to shed light on the populations of chlorophylls that exist in the complexes, and the sequence in which excitation energy passes through them (see Section 2.4), and secondly, changes in fluorescence in the green-plant PSII are indicators of the functioning of the very early stages of electron transfer reactions in the reaction centre itself (see Chapter 5).

Energy transmission

Energy can be transferred from an excited pigment to an adjacent unexcited one in accordance with the principles of 'inductive resonance' described by Förster. Transfer is facilitated by (i) close range: the rate of transfer is inversely proportional to the inverse sixth power of the intermolecular distance, (ii) an orientation such that the directions of the absorption dipole moments are parallel: the rate of transfer is proportional to $\cos^2\theta$ where θ is the angle between them, and (iii) the fluorescence spectrum of the excited molecule having a good degree of overlap with the absorption spectrum of the receiving molecule. It is now clear why chlorophylls are not found free in solution in the lipid phase of the membrane. Because of statistical variation, it would often happen that an excited molecule would be unable to find a partner before it underwent either fluorescence or deactivation. As things are, the chlorophylls being fixed evenly and densely to an extended framework of proteins, the efficiency of transfer is fixed. It is fixed, moreover, at a value close to the optimum, since the loss of energy in the chlorophyll system is very small (as shown by the value of the quantum requirement for oxygen evolution being only just over eight per mole, in adapted algae).

It is also clear why it is found (or at least expected) that the type and binding of accessory pigments to the protein framework is such as to position those with the shortest wavelengths of maximum absorption at the greatest distance from the reaction centre, with a graded sequence leading to the Pλ active centre. Because of Stokes' shift, the fluorescence of a particular chlorophyll will overlap better with another molecule that absorbs further to the red than with one that absorbs further to the blue. Therefore transfer will tend statistically to take place in the direction of increasing wavelength, that is, of decreasing energy of absorption. The difference in energy at each step of course results in a small degree of heating, and in any case the differences are so small that it is estimated that over 1000 transfers take place on average between an absorption event and the excitation of reaction-centre chlorophyll, even though there may be only 200–400 chlorophylls per reaction centre. This is known as a 'random walk' process.

There is a second process that must be considered, that of exciton transfer. This is established for crystals such as germanium. The orbit of the excited electron is large enough to cover neighbouring atoms, and the 'hole' or positive charge centred on the excited atom can be filled by a transfer of an electron from a neighbour. In this way the electron-hole pair, like an atom of 'positronium', can migrate through the crystal. The transfers are short-range, and work below 2 nm, whereas inductive resonance works at 8–10 nm or further. The reaction centre of purple bacteria, as shown by X-ray crystallography, contains bacteriochlorophyll and bacteriophaeophytin molecules which are considerably closer to each other than 2 nm, and the delocalised electron provides a pictorial model of the initiation of the electron transfer that is the primary step in photosynthesis. There are, however, difficulties in applying exciton theory to photosynthetic systems. In no sense could the chlorophyll array in the reaction centre be considered crystalline; the pigment molecules are not identical and the spacings, while precise, have only an approximate symmetry.

Thermal deactivation

Thermal or non-radiative deactivation describes any process whereby an excited molecule returns to its electronic ground state, the excitation energy appearing as heat. Usually a neighbouring molecule is involved. This is likely to be a solvent molecule or another part of the same chlorophyll which happens to be in a state to interfere with the excited state. The lifetime of the excited state is so short that there is no likelihood of a solute molecule approaching and reacting. It appears that deactivation is well controlled by the organisation of the chlorophyll, so that it is negligible in relation to photochemistry. However, the very short lifetime of fluorescence in green-plant PSI is considered to be due to deactivation by or association with the oxidised reaction-centre chlorophyll $P700^+$. Until it is re-reduced (by plastocyanin, see Chapter 6) it cannot use energy, but deactivates it by an unknown mechanism. A second deactivation system is the 'valve reaction' of Wolff and Witt (40) in which very bright flashes produce a transient spectral change in the carotenoids, implying that these are able to collect surplus excitation and deactivate it.

Photochemistry

The excited state as described above is a 'singlet' in the terminology of spectroscopists because the electron that has been promoted is still paired with the spin of the other electron in the original energy level. A magnetic field does not split the band. There is a process whereby the electron can change its spin, becoming parallel with its previous partner, and this state is known as a triplet. It is a 'forbidden' process, but its low probability has to be multiplied by the large number of opportunities.

Virtually all photochemical reactions that take place in solution do so either by the immediate breakage of the absorbing bond, or by means of a triplet state. This is because the triplet state has a much longer lifetime, of the order of 100 μs for chlorophylls, which allows for the movement of molecules. Chlorophyll in solution can be photoreduced to a pink compound by ascorbic acid (the Krasnowskii reaction (41)), but such reactions cannot be models for the photosynthetic reaction centre.

The photosynthetic process has more in common with solid state processes. The only chemical process which can be completed in the nanosecond time scale is the movement of an electron, and a distinguishing feature of electron transfers is that they proceed more readily at cryogenic temperatures than at 20 °C. The essential feature of chlorophyll-based photosynthesis is the sequence containing a photoionisation step:

$$[CX] \xrightarrow{h\nu} [C^*X] \rightarrow [C^+X^-]$$

in which the ion-pair is stabilised by formation of a multimolecular charge-separated state, so that the lifetime is for temperature-dependent processes to operate (this is more fully described in Chapter 3).

2.4 Organisation of Chlorophyll

The association of Chl a and b with polypeptides and protein complexes has been described above. It has also been implied that the organisation of proteins and their subunits has been such as to provide a path whereby excitation appearing in the bulk, shorter wavelength-absorbing chlorophyll was transferred by means of appropriate juxtapositions to an eventual trap at P680 or P700. It is hard to prove this hypothesis. Isolated chlorophyll–polypeptide complexes may be altered slightly, affecting the environment for the chlorophyll, and hence the absorption maxima, and there is no firm basis for judging how the antenna peptides fit around the reaction centre. In this section we have to consider what information can be obtained from studying the optical properties of chlorophyll in relatively intact systems, and how far the belief in a directed path can be justified.

The absorption spectrum of thylakoids can be compared with chlorophyll in dilute solution. (Intact chloroplasts are unsuitable because they scatter light selectively, and also suffer from the flattening of the peak absorption, owing to the chlorophyll at the back of a chloroplast being shaded by that at the front.) The spectrum of the 680 nm absorption of thylakoids is obviously lumpy, not a smooth single gaussian curve. The irregularities in the thylakoid curve can be identified as discrete forms of chlorophyll by, for example, subjecting the curve to mathematical differentiation, which can be done instrumentally; some modern spectrophotometers can plot the first, second or fourth derivative spectra. This is illustrated in Fig 2.13. Alternatively a computer can be used to build up an absorption spectrum from gaussian components, matching the experimental spectrum by trial and error while varying the position of the maxima, the bandwidth and amplitude of each component. An example is shown in Fig. 2.14. The noteworthy result is that there is not a continuous spread of spectral types of chlorophyll, but that only four are found. These types are of course purely physical manifestations of their particular environments. As with the bacteriochlorophyll notation, the forms of chlorophyll in vivo are known as C670, C680, C695 and C705.

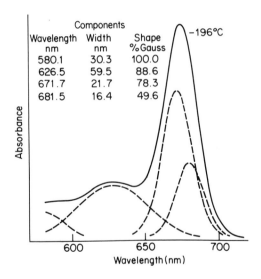

Figure 2.13. The absorption spectrum of a pale mutant of *Chlorella*, from which a computer program has extracted the 'components' as shown. Reproduced from French & Prager (42) with permission

There is interest in studying the orientation of chlorophyll. The maximum absorbance in the red corresponds to a dipole in the molecule extending approximately along the axis of the rings I and III. This is known as the y-axis, and the absorption at the red maximum is ascribed to the 'Q_y transition'. The average direction of the y-axes can be determined in thylakoids that are oriented, which can be done by means of a magnetic field, or by spreading the membranes on

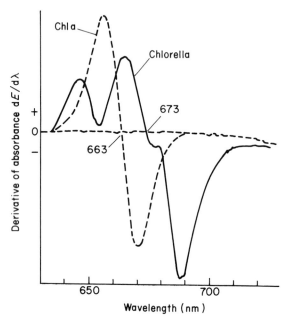

Figure 2.14.   The derivative of the absorption spectra of Chl a in ether solution (dashed line) and of *Chlorella* (continuous line). The presence of Chl b at 650 nm, and of Chl a forms at 670 and 684 nm, are revealed in *Chlorella*. Higher derivatives, up to the fourth, are often used. Reproduced from French & Huang (43) with permission

a glass plate, or by casting them into a flexible gel and then deforming the gel. The conclusion reached was that the shorter-wavelength forms (the majority of the chlorophyll) were inclined at an angle to the plane of the membrane, but the long-wavelength forms were closer to the plane of the membrane. The same conclusion holds with protein–chlorophyll complexes such as aggregated LHCII, PSI and PSII particles, assuming that in the membrane the long axes of the particles are in the plane of the membrane.

An answer to the question of alterations being introduced into the pigment arrangement when peptides are isolated is given by the technique of circular dichroism. This measures only the absorbance of chromophores, in this case chlorophyll, where there is appreciable chirality (or asymmetry). The technique is powerful for simple molecules, but with several chlorophylls in a closely interacting group the spectra are hard to interpret. Nevertheless, the spectra are very sensitive to changes in the organisation of the chlorophyll, and the apparent identity between the CD of thylakoids and the sum of the CD of the complexes derived from them (44) indicates that there has been no alteration, during the protein purification, in the relative positions of (at least) those chlorophylls that are responsible for the CD.

Fluorescence studies

Some of the most far-reaching concepts of chlorophyll organisation have come from observations of fluorescence. This is a technique with a long history. It was clear from work in the early 1960s that fluorescence from chloroplasts was heterogeneous, and measurements of emission spectra, lifetimes and the relative intensities of the different components have been continuously and vigorously pursued, notably by the late Warren Butler.

At room temperature, chloroplasts illuminated at 435 nm fluoresce with a major emission peak at 685 nm, with sometimes a shoulder near 695 nm and a minor broad peak at 720–740 nm (Fig 2.15). At low temperature (77 K) the main components are at 730 and 720 nm, and there are clear peaks at 695 and 685 nm. These peaks persist when the chloroplast is resolved into PSI and PSII particles.

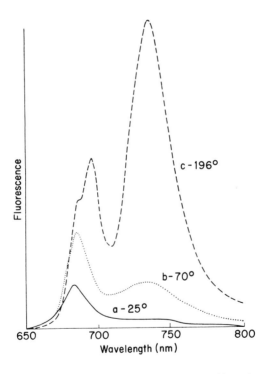

Figure 2.15. Fluorescence emission spectra of spinach chloroplasts. Spectra were recorded with a 5 nm half bandwidth, with excitation at 475 nm. The lowering of the temperature from $-25°C$ (a) to $-70°C$ (b) and $-196°C$ (c) produces a great increase in the fluorescence of PSI. Reproduced by permission of Professor Norio Murata and Elsevier Science Publishers. From Murata, N., Nishimura, M. & Takamiya, A. (1966) *Biochim. Biophys. Acta* **126**: 237

PSII particles retain the emission at 685 nm. (The fraction of the chlorophyll responsible for the emission is labelled F685.) F685 originates from C680 (Chl a absorbing at 677–680 nm). There is little or no fluorescence from the LHCII when it is attached to PSII, showing that CCII is an effective energy trap: isolated LHCII has a fluorescence visible to the eye. LHCII contains C670. C680 accounts for most of the chlorophyll in PSII; it is not at present possible to say how the different chlorophyll groups are distributed between the pigment–protein complexes CP47, CP43, CP29 and the D-proteins.

PSI particles emit very much less fluorescence than PSII at 298 K. Preparations containing 100 or more Chl a per P700 show an emission maximum at 685–690 nm, with only a shoulder at 700-740 nm. The shoulder is composed of the vibrational tail of the 690 nm emission, with contributions from minor components at 720 and 735 nm. At 77 K, however, the maximum is at 735 nm, with a major sub-component at 720 nm. The 690 nm component remains, but is relatively insignificant. At low temperatures the fluorescence of PSI is thus greatly increased at 720 and 730 nm, and is greater than that of PSII at 685 nm.

In Section 2.2 the two antenna complexes, LHCIa and LHCIb, were described. LHCIa emits fluorescence at 690 nm, LHCIb at 730 nm (27). The core complex CCI contains four groups of Chl a molecules: P700 and the pigments immediately associated with it as electron carriers (these are too small a minority to play any part in detectable fluorescence), a 'core antenna' containing perhaps 40 tightly bound chlorophylls, an 'internal antenna' (20–25 Chl) and a 'peripheral antenna' (40–45 Chl) (46). All three antennae in CCI contain C680 (F690); the internal antenna contains in addition C697 (F720) and the peripheral antenna (in higher plants) contains C705 (F735). These assignments have been made on the basis of the ease of removal of chlorophyll from CCI; thus removal of some 40 chlorophylls removes F735, etc. As in the case of PSII, it is not possible to associate these antennae with specific protein complexes in CCI; there are in any case only two polypeptides known to carry chlorophyll, CP71 and CP67. PSI may be tri- or polymeric in situ, and the more loosely bound chlorophyll may be carried between polypeptides, as opposed to other chlorophyll bound entirely within one polypeptide.

Bassi and Simpson (27) were able to show that LHCIb, when attached to CCI, quenched the 720 nm fluorescence from the (internal antenna of) CCI. They concluded that LHCIb is physically closest to CCI. They also showed that LHCII could complex with PSI, and pass energy to it, but only when LHCIa was present. Hence LHCIa is on the outside of PSI, and PSI can at least in principle attach to LHCII. Their model shown in Fig. 2.11 is a quantified extension of the earlier one (24).

In addition to the emission spectrum and relative intensities, the fluorescence components can also be characterised by their lifetime, which can now be measured by means of the single-photon counting technique. A relatively weak pulse of about 10 ps duration, from a laser, excites the sample, and the equipment measures the time for the arrival of the first fluorescent photon in a given wavelength interval. The pulses are rapidly repeated, and the sample is circulated so that there is no actinic effect. (An actinic effect is a detectable photochemical change brought about by absorbed light. It is necessary to avoid such changes in crucial components such as electron acceptors, and also to avoid the mutual annihilation of two quanta in the same pigment bed, which produces unusually short lifetimes when very high-energy, short-time flashes are used.) The distribution of arrival times is analysed by computer, on the basis that the sample contains a discrete number of components, each of which is excited according to the shape of the exciting pulse, and decays exponentially with a characteristic yield, spectrum and lifetime. The computer finds by trial and error a set of magnitudes and lifetimes that account for the observed decay curve within a preselected margin of error. One criticism of the method is that present computational methods can only handle about four decaying components, and when four are found one may suspect that there could have been five. The computer would find different values for the first four if it could also find a fifth.

The fluorescence emission with the maximum at 685 nm, observed in whole cells and PSII preparations, was resolved by means of the deconvolution procedure into two components with almost identical spectra (46). They differed in their lifetimes (0.2 ns and 0.5 ns with the traps open, and 2 and 1 ns respectively with the traps closed). Their dependence on the state of the traps showed that they were both produced from PSII. When the excitation spectrum was examined, Chl b was found to excite only the first of the two, and this was interpreted as an indication that it alone was coupled to the antenna complex LHCII. It is therefore probable that the two F685 components represent the cores of PSIIα and PSIIβ reaction centres (see Chapter 1). There was also a component at about the same wavelength of emission, with a lifetime of 2.2 ns (independent of trap state). Its intensity was $1/100$ that of the PSIIα and β components, and it was interpreted as arising from LHCII; for practical purposes, LHCPII does not fluoresce in situ.

When the cells (of green algae) or chloroplasts were taken to state II, the antenna size of PSIIβ increased and PSIIα decreased, consistent with a migration of LHCII (after phosphorylation) from PSIIα in the grana to PSIIβ in the stroma thylakoids. However, it has not yet been shown that the fluorescence of PSIIβ in state II is excited by light absorbed by Chl b.

The fluorescence of CCI preparations containing 110 Chl a per P700 was analysed by means of time-resolved fluorescence and one component was also found to have a very short lifetime (reported by various groups as 10–16, 40–60 or 70–100 ps). Its emission wavelength was 690–695 nm (F695), and was more or less invariant with temperature. The appearance of F720 and F735 (only observable at low temperature) was delayed relative to F695 (Fig. 2.16), and had a lifetime of 320 ps. F720 and F735 are believed to account for the 2–3 ns

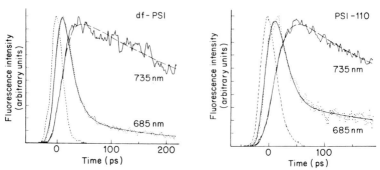

Figure 2.16. Fast kinetics of PSI fluorescence. In each diagram the time-resolved fluorescence at 77 K is shown for a particular type of PSI preparation, df-PSI (detergent-free) and PSI-110 (110 Chl per P700). The exciting pulse of laser light is shown as a dotted curve, and the decay curves for two fluorescence components, extracted by computer, are shown. The components have emission wavelengths of 685 and 735 nm. The 685 nm fluorescence decays with two time constants, 13.7 and 189 ps in df-PSI and 12.1 and 305 ps in PSI-110. The decay of the 735 nm fluorescence had a time constant of 310 ps in df-PSI. The relationship where the maximum rate of increase of one curve corresponds to the maximum height of another indicates a product–precursor relationship. Reprinted by permission of Professor B.P. Wittmershaus and Kluwer Academic Publishers. Copyright © 1986 by Martinus Nijhoff Publishers. From Wittmershaus, B.P. (1987). In Biggins, J., ed., *Progress in Photosynthesis Research*, Proc. VIIth Int. Congr. Photosynth. 1986, vol. 1, p. 78

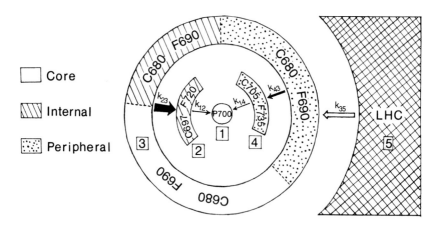

Figure 2.17. Model of fluorescent components of PSI. This concept of PSI differs from that shown in Fig. 2.11, in that the different fluorescent components do not (necessarily) correspond with resolvable chlorophyll–protein complexes. Reprinted by permission of Professor B.P. Wittmershaus and Kluwer Academic Publishers. Copyright © 1986 by Martinus Nijhoff Publishers. From Wittmershaus, B.P. (1987). In Biggins, J., ed., *Progress in Photosynthesis Research*, Proc. VIIth Int. Congr. Photosynth. 1986, vol. 1, p. 78

lifetime of PSI fluorescence of chloroplasts at 77 K. The yield and the lifetime of both F720 and F735 is dependent on the redox state of the acceptors for P700. One model which attempts to make some order from this complicated data is that given by Wittmershaus (46) and shown in Fig. 2.17. This diagram is based on the calculated rates of transfer of energy from one fluorescent group to another, and it shows that the relatively small F720 and F735 pools are essential in the transfer of energy from the F690 pool to P700, irrespective of whether the F690 is in the core antenna or the internal or peripheral antennae. The C680 (F690) population as a whole is excited via LHCI, consistent with Fig. 2.11.

The lifetime of the C680 (F690) is very short because it is quenched (the energy is taken) by C697 (F720) (72%) and by C705 (F735) (28%), shown by the precursor–product relationship in the decay and rise times of the fluorescence components. The F720 and F735 complexes are parallel routes for energy to reach P700. P700 is not an efficient trap with respect to its immediate suppliers because the difference in energy is so small.

At room temperature the fluorescence of the F720 and F735 is quenched, probably by the oxidised reaction centre $P700^{+}$. That is to say that energy arriving in CCI complexes is rapidly converted to heat, as long as the P700 is oxidised. This provides a rationale for segregating the two photosystems.

Further Reading

A general treatment of light absorption, spectroscopy and photochemistry will be found in:

Smith, K.C. (1977) *The Science of Photobiology*, Plenum, New York, chs. 2–3, pp. 27–86.

Chapter 3

Generalised Reactions — Overview

The object of this chapter is to describe some features that are common to the patterns of photosynthesis found in the two photosystems of green plants and in bacteria. This is intended to avoid repetition later (the differences are detailed in Chapters 4–7). However, the consideration of analogous systems and components in the different groups leads on to discussions of homology, and hence of an evolutionary order in the diversity of Nature—an exciting question (even though it may be impossible to answer). Furthermore, if we expect to find unity in Nature, then insights gained into a process in one organism may well be applicable to another, and in this chapter it will be apparent that such insights have been a two-way trade across the notional frontiers of photosynthesis.

3.1 The Generalised Photosynthetic Reaction Centre

The function of a reaction centre is to drive an electron transport system. The common feature of reaction centres is simply stated: there is a chlorophyll molecule or dimer (C) in close proximity to an electron acceptor (X). The proximity is so close that when C becomes excited directly or indirectly by light energy, an electron passes from C to X without the need for any molecular movement. This transfer is very quick (of the order of a few picoseconds) and takes place even at liquid helium temperatures, which requires that C and X are part of the same protein complex.

In all known cases, there is a series of electron acceptors (X_i) in the reaction-centre complex. The electron travels from one to the next, each step taking a longer time and being more temperature-sensitive. The electron finally leaves the reaction-centre complex by reducing a diffusible molecule (X_n). The positive charge left on C is filled by an electron from a donor (D_i) which may or may not be part of the complex. In general the restoration of C^+ to C is relatively slow and temperature-dependent. Ultimately a diffusible donor molecule restores the complex to its original state, so that at this point the effect of the absorption of the quantum of light energy in the complex can be identified as the production of oxidised D_n and reduced X_n, as set out in Fig. 3.1.

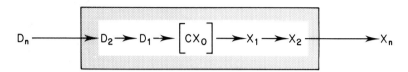

Figure 3.1.   Generalised diagram of a photosynthetic reaction centre. C is a pigment such as chlorophyll, which on excitation transfers an electron to the acceptor X_0, which is in close contact (square brackets). The electron passes along a chain of acceptors, still inside the same protein complex (shading) and reaches the diffusible acceptor X_n. The positive charge left on C is filled by bound donors D_1, D_2 (which may not exist in every case), the ultimate donor being the diffusible compound D_n

Photosynthetic reaction centres are always found in membranes, and it is inconceivable that the separation between the donors and acceptors above could be effectively maintained without putting them into different phases (two aqueous, one lipid), so as to prevent them from back-reacting. Nevertheless all the various types of reaction centre have been isolated and found to function in solution, experimentally; the lipid membrane is not necessary for their photophysical action.

Table 3.1 displays the identities of D_n and X_n for the four types of reaction-centre complex. The details and the evidence for each are covered in Chapters 4–7.

Table 3.1   The four types of photosynthetic reaction centre

Type	Designation	Electron donor D_n	Electron acceptor X_n
Purple bacteria Green non-sulphur bacteria		cytochrome c	ubiquinone
Green sulphur bacteria		cytochrome c-553	ferredoxin
Green plant	PSII	water	plastoquinone
Green plant	PSI	plastocyanin or cytochrome c-553	ferredoxin

The reaction centres may be regarded as enzymes, catalysing the oxidation of the donor and the reduction of the acceptors shown, by means of light energy, with the important proviso that the reverse reaction is effectively prevented.

The electron transfers in the reaction-centre complex are examples of redox reactions. The diffusible products (D_i and X_i) may undergo further redox reactions, for two purposes: the formation of metabolically useful products (e.g. the reduced coenzyme NADPH), and the production of ATP. For this discussion, we require the concept of redox potential.

3.2 Redox Potentials

The electric potential at a given point is the work done, per unit charge, in bringing an additional infinitesimal charge from an infinite distance to that point. For practical purposes we do not deal in free electrons at infinite distances, but in movements of electrons from one point to another, and we are interested in potential differences rather than absolute potentials. When an electron (negatively charged) moves from a negative to a positive potential, the total stored work is diminished, and energy is released, as heat. The process is spontaneous. Movement of an electron from one molecule to another is an example of an oxidation–reduction (redox) reaction. The oxidant gains the electron, the reductant loses it. The acceptor becomes reduced. Other examples are addition of oxygen (oxidation) or of hydrogen atoms (reduction); in many cases the addition of oxygen or hydrogen is found to be preceded by an electron transfer.

Substances, or more precisely couples consisting of two substances that are interconverted by a redox reaction, can be placed in order in a series by determining for each its standard redox potential (E_0, also known as the mid-point potential E_m). This is done, for example, by constructing the electrochemical cell shown in Fig. 3.2. The left-hand half-cell contains a solution of the two substances of the couple (each at 1 M concentration) in a medium of defined pH (pH 0 unless stated) with a platinum electrode, at a defined temperature. If the couple does not readily react at a platinum surface, a suitable sensitiser is added. The right-hand half-cell is a standard hydrogen electrode, in which hydrogen gas at standard pressure is in contact with platinum in the pH 0 solution. A hydrogen electrode is highly inconvenient, and one would normally use a reference electrode (probably built in to the platinum electrode assembly) of Ag–AgCl or $Hg-Hg_2Cl_2$, in potassium chloride solution, which has been standardised with respect to a hydrogen electrode. The solutions in the two half-cells are connected by a 'salt-bridge' containing saturated potassium chloride solution. The potassium and chloride ions diffuse at more or less equal velocities, and effectively swamp the effect of the unequal diffusion of any other ions, thus eliminating the troublesome junction potential. The standard potential of the couple is the potential of the left-hand platinum with respect to the right-hand platinum, measured with a device that does not draw current (a potentiometer).

A given substance may undergo more than one redox reaction, and therefore belong to more than one couple. For each case, there will be a specific standard

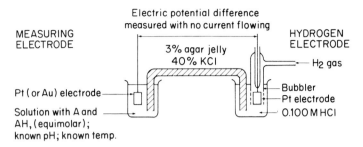

Figure 3.2. Principle of determination of redox potentials. In the apparatus shown, the redox potential of the couple A/AH is obtained with respect to the standard hydrogen electrode. When [A] and [AH] are at unit concentration, the potential is the standard potential for the couple at stated temperature and pH. (Note that to conform with theory all solutions should be at unit activity rather than unit concentration)

potential. A phrase such as 'the standard redox potential of cytochrome c' refers, in shorthand, to the couple: reduced cytochrome c, oxidised cytochrome c.

In the cell of Fig. 3.2, the concentrations of the oxidant and reductant may be varied. The platinum will take a new, non-standard, potential (E) given by equation 3.1:

$$E \ = \ E_0 \ + \ (RT/nF) \ \ln\left((\text{oxidised})^a/(\text{reduced})^b\right) \qquad (3.1)$$

where R is the gas constant, T is the absolute temperature, n is the number of electrons transferred according to the equation of the reaction, F is Faraday's constant, and (oxidised), (reduced) are the concentrations of the two forms of the couple, each raised to the power of the number of molecules represented in the equation. In the commonest cases where one molecule of the oxidised form yields one molecule of the reduced form, the powers a and b are unity, and it is the ratio of concentrations that is important rather than the actual concentrations. The equation can be simplified further by combining the constants, at $25°C$:

$$E \ = \ E_0 \ + \ (59 \text{ mV}/n) \ \log\left((\text{oxidised})/(\text{reduced})\right) \qquad (3.1b)$$

Reactions such as the oxidation of Fe^{2+} to Fe^{3+} only involve an electron. In most redox reactions in organic chemistry, as well as in the couples H_2O/H_2 and O_2/H_2O, hydrogen atoms are involved, and hydrogen ions (H^+) in solution appear in the equation of the half-cell:

$$AH_2 \ \rightarrow \ A \ + \ 2H^+ \ + \ 2e^-$$

In such cases the potential of the couple AH_2/A is pH-dependent. The symbol E'_0 (E'_{0pH}, $E_{m,pH}$) is used to denote the standard potential of an equimolar mixture of AH_2 and A at a stated pH. The pH-dependent potential is given by equation 3.2:

$$E'_0 = E_0 - (59 \text{ mV}/n) \text{ pH} \tag{3.2}$$

(since pH is the negative logarithm to the base 10 of the hydrogen ion concentration). n in equation 3.2 is the number of hydrogen ions accompanying the transfer of one electron. The importance of this concept is shown in Fig. 3.3, in which E'_0 for a quinone is shown to have three different dependencies on pH, according to whether it is undissociated, singly ionised or doubly ionised. It should also be noted that the hydrogen electrode potential is only defined as zero at pH 0; at pH 7 it has a potential of -0.42 V. The couple O_2/H_2O has a potential of $+0.815$ V at pH 7. Obviously there is little biological value in working at pH 0, even when the materials are able to tolerate it.

Redox potentials are additive; in the cell of Fig. 3.2, the potential measured for couple A against the hydrogen electrode, minus the potential for couple B against the hydrogen electrode, is equal to that of couple A measured against couple B.

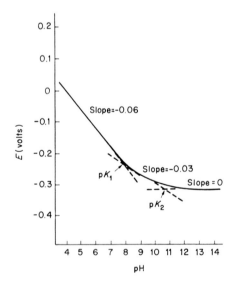

Figure 3.3. Dependence of standard redox potentials on pH. The graph shows the standard redox potential of anthraquinone 2,7 disulphonate as a function of pH. Below the lower pK value, the reduced form is the quinol; two protons are taken up per two electrons. Between the pK values, only one proton is taken up per pair of electrons, and above the higher pK, the potential is invariant with pH. Reproduced from Mahler & Cordes (47) with permission

One of the uses of a knowledge of standard redox potentials is to enable an investigator, knowing what materials are present in a system, to predict the order in which they react. This has been of the greatest importance in the early discussions of both mitochondrial and photosynthetic electron transport. The components are placed in such an order that with reasonable proportions of the oxidised and reduced forms present, there is a steady progression of potential from negative to positive. When the standard potentials of two or more components are very similar, however, say within 59 mV for one-electron processes, the method is not able to predict the sequence because an adjustment of the relative concentrations within the range 1 : 10 to 10 : 1 would satisfy the need for progression in any order.

3.3 Electron Transport

An electron moves up a potential gradient, that is, from a relatively negative to a relatively positive potential. The potentials are of course not the standard potentials, but those given by the concentration ratios of the oxidised and reduced members of each of the couples.

3.3.1 Driving redox reactions with light energy

The importance of chlorophyll as a driving force for electron transport can be seen from the following data of Seely (48). Chlorophyll a (C) can be oxidised and reduced, and standard potentials are known:

$$C + e^- \rightarrow C^- - 870 \text{ mV} \qquad (i)$$

$$C \rightarrow C^+ + e^- + 780 \text{ mV} \qquad (ii)$$

$$C^* \rightarrow C^+ + e^- - 1070 \text{ mV} \qquad (iii)$$

$$C^* + e^- \rightarrow C^- + 980 \text{ mV} \qquad (iv)$$

Process (i) expresses the difficulty of reducing chlorophyll; C is a very poor oxidising agent, but C^- is a very powerful reducing agent. Likewise, from (ii), C is a poor reducing agent, but C^+ is a very good oxidising agent. Processes (iii) and (iv) show that C^* is both a very good reducing agent and a very good oxidising agent. (C^* oxidises and reduces itself when it fluoresces.)

If the process is imagined whereby reactions (ii) and (iii) are combined

$$C^* \rightarrow C^+ + e^- \rightarrow C \qquad (v)$$

an electron has in effect gone from couple (iii) to couple (ii), from -1070 mV to $+780$ mV. This releases 1.85 eV of energy (as heat), and accounts for the energy of the photon of light of 670 nm that effected the transition

$$C \overset{1.85 \text{ eV}}{\longrightarrow} C*$$

(more strictly, the zero-zero energy shown in Fig. 2.12). It would of course be equally possible for photosynthetic systems to use processes (i) and (iv) instead of (ii) and (iii), but they do not. Process (v) becomes a useful photosynthetic electron transport system when the electron is required to pass along a chain of carriers before it is allowed to recombine with C^+. We now consider the general features of such chains.

3.3.2 The redox components of photosynthesis

Quinones

Ubiquinone (UQ) in most bacteria and mitochondria, menaquinone (MK) in other bacteria and plastoquinone (PQ) in green plants are the diffusible molecules found dissolved in the lipid structure of the membranes. They have similarities in structure such as the isoprenoid tail that confers lipid solubility (formulae are given in Fig. 3.4). They are reduced by the reaction centres of green-plant PSII and purple bacteria (and by metabolic substrates via membrane-bound reductases in mitochondria). Their reduced forms are known as quinols, and are hydrogen carriers. This enables them to play a

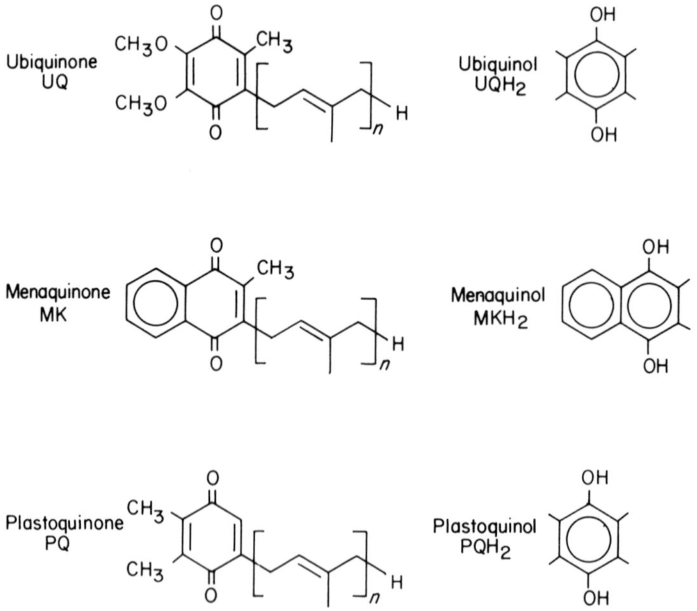

Figure 3.4.   Quinone/quinol couples of importance in biological electron transport

major role in the pumping of hydrogen ions from one side of a biological membrane to the other, generating the energy store (PMF) that ultimately provides for ATP formation. When quinones are reduced to quinols they acquire hydrogen ions from the aqueous P phase, as required by the equation

$$Q + 2e^- + 2H^+ \rightarrow QH_2$$

When the quinols are reoxidised to quinones by the next step of the electron transport chain, the protons are released on the E side:

$$QH_2 \rightarrow Q + 2e^- + 2H^+$$

Cytochromes

Cytochromes are proteins which carry haem groups in which the central iron atom is reversibly oxidised between Fe^{2+} and Fe^{3+} states. The *a*-cytochromes, which have an open coordination position on the iron available for oxygen binding, are not represented in photosynthesis. The *b*-cytochromes are characterised by a protohaem IX prosthetic group coordinated to two protein histidine residues (the haem can be dissociated from the protein in acid acetone). The *c*-cytochromes, on the other hand, have additional thioether bonds from the 2- and 4-positions of the tetrapyrrole ring to protein cysteine residues. Cytochrome *f* (in chloroplasts) is a *c*-type cytochrome (c_6). The *b*-cytochromes in photosynthetic systems are always part of protein complexes in the membrane, but cytochromes *c* and c_2 are found in mitochondria and bacteria as diffusible molecules in the E space. Table 3.2 summarises the properties of redox proteins in photosynthesis.

Plastocyanin, a copper protein

Plastocyanin in thylakoids of chloroplasts and cyanobacteria appears to carry out the role of cytochromes *c* and c_2 in that it is a diffusible carrier in the E space, that is, in the lumen of the thylakoid. It is specific to PSI and is described in Chapter 6.

Iron–sulphur proteins

The first members of this group to be discovered were the ferredoxins: soluble, small proteins with either Fe_2S_2 or Fe_4S_4 clusters. The former were pursued by R. Hill as the 'met factor' which enabled chloroplast thylakoids to reduce metmyoglobin (the role of NADP not being known). A. San Pietro isolated the protein, as PPNR (phosphopyridine nucleotide reductase, a false identification), and it was named ferredoxin by D.I. Arnon on account of its similarity to

Table 3.2 Properties of some redox proteins in photosynthesis

Protein	Prosthetic group	Abs. max. (nm)	Molecular mass (kDa)	Redox potential (mV)	Location
cytochrome:					
b	haem (coordinate)	560*	40	−60 to −90 and +50*	bc-complexes in bacteria
b_6		563	24+15	−170 and −50	b_6f in chloroplasts
b-559		559	9+4	370 or 65	plant PSII
c	haem (covalent)	550	12.5	254*	periplasmic bacteria
c_2		555	13.5	300	
$f (= c_6)$		554	35	340	b_6f in chloroplasts
c-555,553		555,553	36	350 and 130*	reaction centre of purple bacteria
c-552		552	12	370	thylakoid lumen in *Euglena*
plastocyanin	Cu	598	10.5	370	thylakoid lumen
ferredoxin (plant)	Fe_2S_2	420	12	−420	chloroplast stroma
(bacterial)	Fe_4S_4	420	6	−450	sulphur bacteria
ferredoxin:NADP reductase	FAD	455	44	−380?	thylakoid P-surface

Absorption maxima are given for α-bands of reduced cytochromes, and for oxidised states of other proteins. Differences between species must be allowed for; asterisks indicate wide variation.

the second group, discovered in the non-photosynthetic bacterium *Clostridium pasteurianum* (49). The Fe_2S_2 ferredoxins are one-electron carriers, and alternate between Fe^{3+}–Fe^{2+} and Fe^{2+}–Fe^{2+} states. Bacterial types with four iron atoms can carry one or two electrons. The structure of the clusters is shown in Fig. 3.5. Both types of ferredoxin have very negative redox potentials, equivalent to the hydrogen electrode at pH 7 ($E_{m,7} = -0.42$ V). Similar low-potential iron–sulphur proteins occur bound in the PSI complex (see Chapter 6).

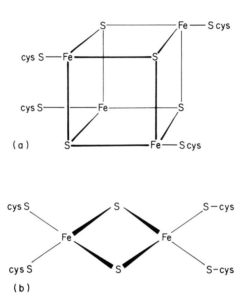

Figure 3.5. Ferredoxins (iron–sulphur proteins). (a) Bacterial type (Fe_4S_4). (b) Plant type (Fe_2S_2)

Another group of iron–sulphur proteins occurs in the ubiquitous redox particle known as the bc_1 or b_6f complex, found in the membranes of bacteria and mitochondria, and chloroplasts, respectively (see below). These are the Rieske proteins first isolated from complex III (the bc complex) of mitochondria (50). They have a characteristic ESR (electron spin resonance) spectrum, and potentials in the region of 0.2–0.3 V. In the purple bacteria the visible absorption spectrum of the Rieske centre was identified as cytochrome cc' or high-potential iron protein (HiPIP).

Cytochrome f, cytochrome b_6 and subunit IV (which are together homologous with mitochondrial cytochrome b) are coded and synthesised in the chloroplast, while the Rieske protein and subunit V are coded in the nucleus and synthesised in the cytoplasm (see Table 13.3). All four major protein complexes in the thylakoid membrane have a similar heterogeneous origin.

Flavoproteins

The enzyme ferredoxin:NADP reductase (FNR) carries one molecule of flavin adenine dinucleotide (Fig. 3.6). This group carries two atoms of hydrogen, which are probably transferred as one proton and one hydride ion. It is specific to PSI and the green sulphur bacteria (see Chapters 6 and 7).

Other FAD and FMN enzymes occur in the membranes of photosynthetic bacteria, in complexes that may be thought of as analogous to the systems NADH:ubiquinone reductase and succinate:ubiquinone reductase of mitochondria. These may be required (photosynthetically) for reverse electron transport (see Chapter 5).

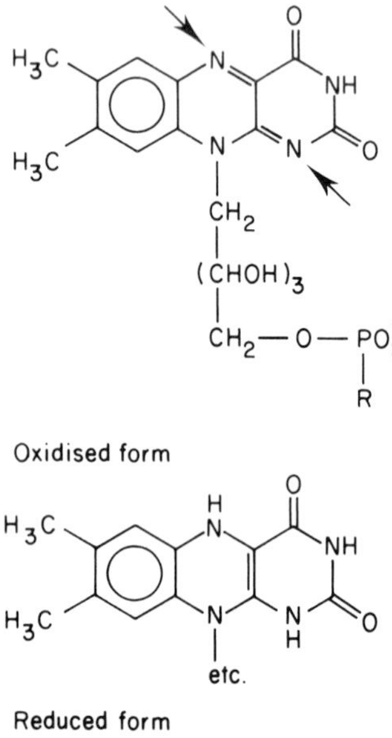

Oxidised form

Reduced form

Figure 3.6. Flavin coenzymes. The role of the isoalloxazine ring in carrying two atoms of hydrogen (in the reduced form) is shown (arrows). R = OH, flavin mononucleotide; R = AMP, flavin adenine dinucleotide

3.3.3 The cytochrome *bc* complex

There is a component of electron transport involving a non-photochemical complex known as the cytochrome *bc* or b_6f complex which may be universal in organisms or organelles carrying out energy-conserving (i.e. ATP-forming) electron transport processes. The mitochondrial version 'complex III' or bc_1 particle was isolated by Green and Wharton (51) by means of cholate detergents, and the chloroplast 'b_6f' particle first appeared as a pink band from a density-gradient centrifugation when Wessels (52) fragmented chloroplasts by means of digitonin. The significance of the association of the two cytochromes (b_6 and f) was not realised until much later. Apart from small differences in the size of the proteins present and their amino acid sequences, and small differences in the redox potentials of the electron transport carriers, all the bc complexes purified to date have contained one molecule of cytochrome b (containing two differentiable haems with standard potentials close to zero), one molecule of bound quinone (potential approximately 0.1 V), one molecule of cytochrome c_1 (potential 0.2 V) and the high-potential iron–sulphur cluster (Fe_4S_4, potential 0.25–0.29 V) identified as the Rieske centre. (It is possible that the above list represents a monomer, and that the actual particle is dimeric, but studies which measured the area of the complex presented as a target for radiation confirmed that the normal particle is in fact monomeric (53).)

The bc particles are intrinsic membrane protein complexes; the diffusible quinone substrate is present in the membrane, and the cytochrome c oxidant is either extrinsic on the E surface of the membrane or (more likely) in free solution in the aqueous E phase (which is the periplasmic space in bacteria, or the inside of chromatophores). All the bc complexes may be regarded as enzymes, catalysing the reduction of cytochrome c, c_2 or c-552 and the oxidation of plastoquinone or ubiquinone.

There is less similarity in their actual structure, notably in that mitochondria have ten polypeptides compared to four (major) in bf and five in bacterial bc_1 complexes. Mitochondrial b has a mass of 42 kDa, and corresponds in sequence homology to two chloroplast peptides: b_6 (corresponding to the N-terminal portion of b), 23.7 kDa, and a 15.2 kDa peptide that corresponds to the C-terminal end, but has no obvious function. The amino acid sequences suggest that mitochondrial b has nine membrane-spanning α-helical portions, of which five and three respectively are represented in the chloroplast b_6, one being lost. Cytochrome b in *Rps sphaeroides* is a 40 kDa peptide. Although cytochrome f matches c_1 in mass (34 kDa), the amino acid sequences show no obvious homology, although they both have the one membrane-spanning sequence. The mitochondrial Rieske iron–sulphur protein is larger than those in the photosynthetic organisms, and they have the one membrane-spanning sequence. The remaining peptides in bacteria, 10 and 8 kDa, have no obvious function, likewise the two small (less than 6 kDa) peptides from the b_6f particle. In addition to the small peptides, mitochondria have two large core peptides. The enzyme FNR (see above) is usually found attached to the b_6f complex, but it can be removed without impairing the normal activity of the complex, and is not therefore truly part of it. However, FNR may react directly with the complex (see Chapter 6).

The operation of the bc_1 or b_6f complex is complicated and controversial, and what follows is a simplified account. The substrate, quinol (UQH_2, MKH_2, PQH_2) is oxidised to quinone at a site 'Q_z' or 'Q_o', and one electron reduces the Rieske centre, the other passing to the two cytochrome b haems in series and thence to a quinone, tightly bound at site 'Q_c' (or 'Q_r'). The hydrogen ions from the PQH_2 pass to the E aqueous phase. The Q_z site can be inhibited by the quinone analogue DBMIB, but there is a difference between the bc complexes and the bf complex in that, of two powerful inhibitors that attack the Q_c site of the mitochondrial and bacterial complexes, NQNO works poorly in the chloroplast particle and antimycin A not at all.

The reduced Rieske centre reduces cytochrome f or c_1 as the case may be; the cytochrome projects from the membrane on the E side and reacts with the soluble protein carrier plastocyanin or cytochrome c respectively. The other electron from the original PQH_2 at the Q_z site (at the semiquinone level) passes through the two b-haem groups in series, across the membrane to the Q_c site, and reduces the bound quinone to quinol, taking up two H^+ from the P aqueous phase. The quinol exchanges with free quinone at the Q_c site. The released quinol reacts as did the first (substrate) quinol, sending another electron to plastocyanin and another to the Q_c site. This cyclic process (Fig. 3.7) should drive an extra proton from the P to the E side of the membrane for each electron transferred from free UQH_2 to plastocyanin; 0.7 to 0.9 extra protons have indeed been observed (55). The principle of a cyclic process was first proposed by Mitchell as the 'Q-cycle' (56) and has since undergone several variations.

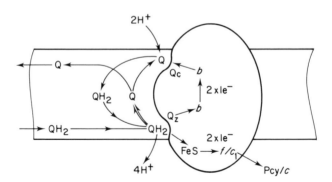

Figure 3.7. A form of the Q-cycle. The cytochrome bc_1 particle is shown. It carries out the reaction catalysed by QH_2: cytochrome c reductase. The complicated path is needed to explain the four protons transferred per pair of electrons transported. Once a QH_2 molecule has given up one electron to FeS, it becomes a semiquinone with a lower potential, able to reduce cytochrome b. Electrons arriving at the bound quinone reduce it to QH_2 which exchanges with Q produced by the other site. Each QH_2 molecule arriving at the complex effectively reacts twice before returning as Q. Some system such as this is well supported by evidence in mitochondria and bacteria, but not at all well in chloroplasts except possibly under low light conditions

The most impressive piece of evidence for a Q-cycle of the above kind is the phenomenon of oxidant-induced reduction of cytochrome b: in the presence of antimycin A, which prevents reoxidation of reduced cytochrome b, addition of an oxidant such as potassium ferricyanide ($K_3Fe(CN)_6$) to mitochondria oxidises cytochrome c_1 and reduces cytochrome b. This is unexpected since the b-cytochrome has the lower potential. The interpretation is that oxidation of the c-cytochrome has oxidised the Rieske centre, which in turn has oxidised a molecule of PQ (by one electron) to a semiquinone, which then reduced the b-cytochrome; the b-cytochrome could not be reoxidised because of the inhibitor.

The discussion so far has assumed that the reductant of the bc complex is a quinol. This may not necessarily be the case in the PSI cyclic system (see Chapter 6), in which reduced ferredoxin reduces the bf complex via an unknown enzyme that is sensitive to antimycin A (ferredoxin:quinone reductase, FQR (57)). (A minority pathway has been reported that does not need soluble ferredoxin, presumably using a bound form of it, and which is not inhibited by antimycin A (58).) These enzymes are distinct from ferredoxin:NADP reductase, a flavoprotein that is almost always found, perhaps as a contaminant, in preparations of the b_6f complex.

The components of the bc and bf complexes are closely analogous except for the cytochromes c_1 and f which have largely defied attempts to find sequence homologies (54). The soluble carriers cytochrome c and plastocyanin are also non-homologous, not only with respect to their prosthetic groups but also in that they are positively and negatively charged respectively, and hence their interaction sites on cytochromes c_1 and f can be expected to be substantially different.

3.4 The ATPase and Photophosphorylation

When the quinols are oxidised by the bc complex the electrons enter the complex but the hydrogen ions are ejected from the membrane into the aqueous E phase. This process therefore results in a membrane potential appearing with the P phase negative with respect to the E phase. In most photosynthetic examples, but not mitochondria, compensating ionic movements take place, which may be either movement in the same direction of anions such as chloride, or movement in the opposite direction of cations such as potassium or magnesium. In such cases the membrane potential is somewhat suppressed, and a pH difference develops across the membrane.

The combined effect of the pH difference and the membrane potential is to attract hydrogen ions back towards the P phase; hence the use of the term protonmotive force (PMF). There is, however, no means by which hydrogen ions can cross the membrane, except via a structure known as the F-ATPase (ATP synthase, formerly called the F_0F_1-ATPase), which couples the movement of three H^+ to the formation of one molecule of ATP. The protonmotive force is a powerful concept introduced by Mitchell by analogy with electromotive force, and has units of volts. Alternatively the electrochemical potential (for protons) can be used, which has units of kJ mol^{-1}. A recent discussion of these terms (59) recommended the symbol Δp, in units of volts; these authors pointed out that the algebraic sign of Δp (etc.) was frequently confused, and proposed that the sign should be that of the inside of a vesicle (meaning a P space), negative when protons are attracted in from the outside. The present author would prefer to specify P space directly, since thylakoids and chromatophores are vesicles in which the insides are E spaces.

$$\Delta \mu_{\mathrm{H}} + \ = \ -5.7 \ \times \ \delta \mathrm{pH} \ + \ E_{\mathrm{m}} \ \times \ 96.5 \text{ units in kJ per mole } \mathrm{H}^{+}$$

$$\mathrm{PMF} \ = \ \Delta \mathrm{p} \ = \ -5.7 \ \times \ \delta \mathrm{pH}/96.5 \ + \ E_{\mathrm{m}} \text{ units in volts}$$

When $\Delta \mu_{\mathrm{H}+}$ reaches a value of approximately 20 kJ per mole H^{+}, or $\Delta \mathrm{p}$ reaches some 200 mV, hydrogen ions flow through the ATPase from the E side to the P side (completing a proton circuit) and forming ATP (provided ADP and inorganic phosphate are present), under expected conditions where the free energy of formation of ATP is +52 kJ mol^{-1}.

The coupling in normal membranes between electron transport and phosphorylation not only means that the first causes the second, but also that electron transport is retarded by the PMF. The exact mechanism of the retardation may be disputed, but it has been shown (60) that the fall to pH 5 in the thylakoid lumen slows the rate of reaction between PQH_2 and the b_6f complex, and comparable effects have been claimed in the oxygen-evolving system of PSII, and in the reaction of plastocyanin with PSI. Certain reagents known as uncouplers have the ability to collapse the PMF and hence allow electron transport to proceed at its maximum (unless the light is limiting) while ATP formation is prevented. This is a different effect from that of inhibitors of phosphorylation such as the antibiotics oligomycin or Dio-9, which bind to F_1, preventing phosphorylation but not relieving the PMF. Many uncouplers have been found to be weak acids with appreciable solubility in the membrane in both associated and anionic forms. They are therefore able to shuttle protons across the membrane. It should be noted that in collapsing the PMF, they do not necessarily eliminate the membrane potential or the pH difference; as long as the effect of the two components is equal and opposite the (net) PMF will be zero.

Amines and ammonium salts at millimolar concentration appear to act as uncouplers, but in this case the effect is to accumulate RNH_3^{+} in the lumen of the thylakoid, substituting an ammonium gradient for the proton gradient. An ammonium gradient cannot be used by the ATPase.

Lastly there are the ionophorous antibiotics such as valinomycin, which makes pores in the membrane through which potassium ions can pass. This has the effect of eliminating the membrane potential, provided potassium ions are present, but not the pH difference. Nigericin, on the other hand, facilitates the electroneutral exchange of hydrogen ions for potassium ions, and so removes the pH difference, leaving the membrane potential intact. These agents allow the determination of whether in a given case phosphorylation depends on the pH difference, the membrane potential or both.

The F-ATPase comprises two parts, F_0 and F_1. F_0 is intrinsic and appears to make a channel for hydrogen ions to pass through the membrane (see Fig. 3.8), since F_0 by itself acts as an uncoupler, that is, it allows the protons to equilibrate across the membrane, preventing ATP formation. F_1, an ATPase enzyme when isolated, is extrinsic and attaches to F_0 forming a characteristic knob structure, visible with negative staining in electron micrographs and known as the Fernandez–Moran particle before its true nature was appreciated. F_1 attaches to F_0 on the P side, and movement of protons (driven by the PMF) from the E to the P side through F_1 causes the synthesis of ATP.

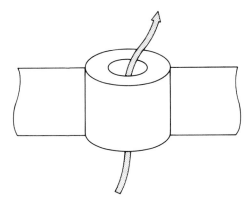

Figure 3.8. The F_0 part of the F-ATPase. The particle forms an annulus composed of several copies each of four types of subunit. By itself, F_0 renders the membrane permeable to hydrogen ions

Four sources have been studied: mitochondria, chloroplast thylakoids, a thermophilic bacterium PS3 and plasma membranes of the bacterium *Escherichia coli*. Their F-ATPases are distinguished as MF, CF, TF and EF respectively. The F_1 all have a common structure of five subunits, $\alpha\beta\gamma\delta\epsilon$ in decreasing order of size. The complex $\alpha_3\beta_3\gamma\delta\epsilon$ (Fig. 3.9) has a mass of the order of 380,000. The F_0 are less well described; four subunits are known, I–IV in chloroplasts, a–d in mitochondria. The polypeptides of both the F_1 and F_0 parts of the ATPase complex have mixed origins; some are coded in the nucleus and synthesised in the cytoplasm, while others are coded and synthesised in the chloroplast. The association of the polypeptides with gene-products is shown in Table 13.4.

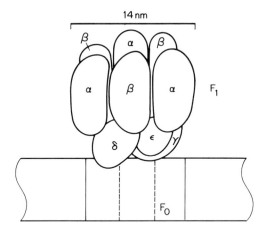

Figure 3.9.   The relation of the F_1 to the F_0 parts of the F-ATPase. Catalytic activity is believed to belong to one or more of the β-subunits. F_1 blocks the hole in F_0 and removes the permeability to hydrogen ions

F_1, the catalytic part of the ATPase, possesses binding sites for ADP and ATP on the α and β subunits. The number of such sites varies. Some sites bind nucleotides very tightly, and since current hypotheses for the mechanism of catalysis predict that the energy-dependent step is the release of bound ATP, the tightly bound sites (on the β subunit) have been seen as catalytic sites. However, ADP and ATP are believed to regulate the enzyme, and therefore regulatory sites must be allowed for.

Chloroplast CF_1 is subject to activation and deactivation, which has been suggested to be an adaptation to prevent the ATPase acting as an ATPase and hydrolysing ATP during the hours of darkness. Activation in the thylakoid requires illumination (to generate a PMF) and the presence of a thiol reducing agent. The isolated enzyme can be activated experimentally by a variety of ways, the most relevant to the natural process being the reduction of a disulphide group in the γ subunit. This could in principle be achieved by the natural thiol protein thioredoxin.

The mechanism of catalysis is not agreed. Nevertheless a model has recently been suggested by Tran-Anh and Rumberg (61) from kinetic studies of the synthesis and hydrolysis of ATP. They noted the following results. In ATP synthesis:

(i) the rate depends on the product $[P_i][H_{in}^+]^2$, hence two protons are involved in the rate-limiting reaction, and

(ii) the rate depends on the reciprocal of $[H_{out}^+]$, hence loss of H^+ precedes substrate binding;

(iii) in ATP hydrolysis there is inhibition linearly dependent on $[H_{in}^+]$ but independent of [ATP], hence one H^+ is released during the rate-limiting step, and

(iv) the Michaelis–Menten relationship with respect to [ATP] is independent of $[H_{in}^+]$, but K_m is inversely related to $[H_{out}^+]$, hence the binding of ATP is preceded by binding of H^+ from the P side.

Assuming that hydrolysis proceeds by the exact reverse of synthesis, they conclude that ATP synthesis involves the following steps (see Fig. 3.10):

(i) The reaction site E gives up a proton to the P phase ($E^{2+} \rightarrow E^+$);

(ii) P_i^{2-} and $ADPMg^-$ bind to the reaction site (E^+) on the P side, forming $[E^+ADPMg^-P^{2-}]$;

(iii) multiple H^+ bind to a binding site (B) at the F_0F_1-interface;

(iv) electrostatic attraction between E^- and B^+ promotes a conformational change in the $\beta\gamma$ subunits;

(v) two H^+ transfer from B to $[E^+ \cdot ADPMg^- \cdot P^{2-}]$ forming $[E^{2+} \cdot ATPMg^{2+}]$ and H_2O;

(vi) one further H^+ transfers from B to form $[E^{3+} \cdot ATPMg^{2-}]$;

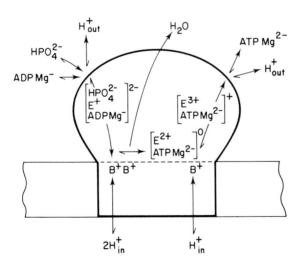

Figure 3.10. A hypothesis for the mechanism of the F-ATPase. The model relates the binding of ADP and P_i to the uptake and loss of hydrogen ions, as required by the kinetic dependence on pH. Reprinted by permission of Professor B. Rumberg and Kluwer Academic Publishers. Copyright © 1986 by Martinus Nijhoff Publishers. From Tran-Anh, T. & Rumberg, B. (1987). In Biggins, J., ed., *Progress in Photosynthesis Research*. Proc. VIIth Int. Cong. Photosynth. 1986, vol. 3, p. 188

(vii) electrostatic repulsion favours the reverse conformational change; and
(viii) $ATPMg^{2-}$ is discharged and E^{3+} releases H^+ to re-form E^{2+}.

3.5 Bacteriorhodopsin

Certain bacteria such as *H. halobium* form patches of cell membrane which
are markedly different from the normal form. These patches are 'purple membrane' and their formation is stimulated in culture by a fall in the oxygenation of
the medium. The membrane consists of lipid and one type of protein only, bacteriorhodopsin (bR), which is a single polypeptide containing 248 amino acids
and forming seven transmembrane α-helical segments. This was established by
means of X-ray crystallography applied directly to purple membrane, since the
bacteriorhodopsin molecules form a densely packed two-dimensional crystalline
array. All-trans retinal is bound as shown in Fig. 3.11 to the lysine-216 residue
by means of a protonated Schiff's base. The absorption spectrum shows a simple maximum at 568 nm. Illumination causes the photocycle illustrated in Fig.
3.12 to take place, beginning with a transient isomerisation about the C13–C14
double bond, so that the chromophore has the 13-cis structure. The proton is
lost from the N of the Schiff's base, and reacquired to complete the cycle. If
the purified bacteriorhodopsin is incorporated into artificial liposomes, it can
be demonstrated that during the photocycle protons are transferred to the inside (E space) from the P space, approximately one to two protons per three
photons absorbed. A potential of up to 300 mV can be formed. If F-ATPase
was also incorporated into the same liposomes, light-dependent phosphorylation
took place. At a time when the chemiosmotic hypothesis of ATP formation was
controversial (with respect to the older chemical-intermediate hypothesis), this
established that at least one organism possessed a plausible method of ATP
synthesis based on hydrogen ion flow, that could be equally plausibly related
to other systems in which F-ATPases were known. This was a timely and ultimately conclusive argument for the validity of the chemiosmotic hypothesis of
ATP production.

It is unlikely that the protons shown in the photocycle, Fig. 3.12 , are in the
bulk phases directly; it is more likely that they are taken up by groups on the
protein. It has been suggested (63) that the cycle can branch, in order to allow
for 'slippage' when the magnitude of the PMF has reached a sufficiently high
value.

Bacteriorhodopsin has a second general point of interest. It was the first protein in which membrane-spanning α-helices were directly proved and located. In
all other cases where proteins are believed to span membranes, the segments
are identified from comparison with the sequences in bacteriorhodopsin. The
procedure is to plot a hydropathy index using a table for the hydropathy of
each kind of amino acid residue, such as that compiled by Kyte and Doolittle
(16).

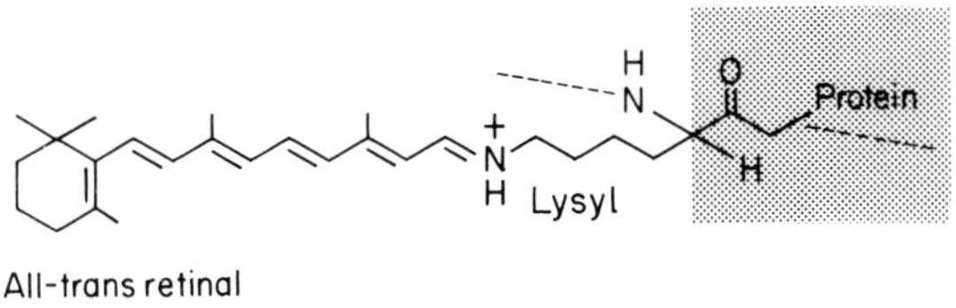

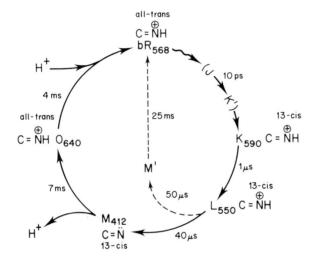

All-trans retinal

Figure 3.11. The chromophore of bacteriorhodopsin. All-trans retinal forms a proton-ated Schiff base with the ε-amino group of lysine-216 in the protein bacterio-opsin

Figure 3.12. The photocycle of bacteriorhodopsin. Conformational and other changes following the absorption light lead in sequence through a series of forms characterised by their absorption maxima and identified by letters (J, K, L, M etc.). Also shown are the likely points of H^+ reaction, and an energy leak via M'. Reproduced by permission of Professor W. Stoeckenius and Elsevier Publications Cambridge. From Stoeckenius, W. (1985) *Trends Biochem. Sci.* **10**: 484

Thirdly, it is arguable that proteins in electron transport complexes, partic-ularly the cytochrome *bc* complex, pump protons as a result of conformational changes. Bacteriorhodopsin may provide a model for such pumping, but even though the structure is relatively precisely known, the conformational changes, if any, are not.

Fourth, the simplicity of the purple membrane has suggested its industrial exploitation for solar energy installations. The efficiency per unit area is not high enough for fuel production, but trials have been carried out of possible de-salination schemes where light is abundant, land cheap and fresh water scarce.

Besides bacteriorhodopsin, *Halobacterium* contains two other photoactive proteins of probably the same general type (see the reviews by Stoeckenius (62,

64)). Halorhodopsin (hR) pumps chloride ions into the cell, and also works by means of a transition between all-trans and 13-cis forms of retinal. There is, however, no protonation of the Schiff's base, and at present there are no viable suggestions for the mechanism of chloride transport. 'Slow rhodopsin' (sR) has a behaviour very similar to hR but 100 times slower, so that it will reach a photostationary state (a particular point of balance between the all-trans and 13-cis forms). It has been argued that a photostationary state provides for responses to light, such as phototaxis. *H. halobium* is a motile bacterium. Most photosynthetic organisms also use light to control growth and development, and the topic of photobiology seeks to provide a common discipline for their study (see, e.g. (65)).

Further Reading

The following text is devoted to the application of the chemiosmotic theory to bioenergetics, and gives a comparative elementary account of the respiratory and photosynthetic electron transport chains, and of the bacteriorhodopsin system:

Nicholls, D.G. (1982) *Bioenergetics: An Introduction to the Chemiosmotic Theory*, Academic Press, London.

Chapter 4

Purple Bacteria

4.1 The Biological Status of the Purple Bacteria

The colour of purple bacteria depends on their complement of carotenoids, and varies from red to brown; some mutants are green. In principle they can be divided into the Thiorhodaceae (purple sulphur bacteria) and the Athiorhodaceae (purple non-sulphur bacteria) depending on whether they use sulphur compounds as reducing agents for photosynthetic carbon dioxide fixation (instead of water as in green plants). These groups are now known as the Spirillaceae and Chromatiaceae, respectively. Some of the Rhodospirillaceae can use hydrogen sulphide (if the concentration is low). All the Chromatiaceae are, however, obligate anaerobes, and oxygen is toxic to them. In this they resemble the green sulphur bacteria. Many of the Rhodospirillaceae are facultative aerobes, and lose most of their photosynthetic pigmented proteins when they are cultured aerobically (which may or may not be in the dark).

The Chromatiaceae typically use hydrogen sulphide to reduce carbon dioxide by the reductive pentose cycle (see Chapter 9). This is photoautotrophic metabolism. Most purple bacteria can use hydrogen gas as an alternative reductant. On the other hand the Rhodospirillaceae normally use organic materials which are elaborated as required using light energy, in other words they live by means of a light-driven reversal of fermentation. This is photoheterotrophic metabolism. The organic materials available may be more reduced or more oxidised than the average of the cell components, and hydrogen gas or carbon dioxide is produced to correct the balance (hence the bubbles from pond mud). However, in several species enzymes can be induced for photoautotrophy using sulphur (or its compounds) or hydrogen as reductant. Furthermore, both groups of purple bacteria can incorporate acetate by photoheterotrophy, and it is apparent that the distinction between them is often difficult. However, relatively few species have been intensively studied.

The purple bacteria and the green non-sulphur bacteria have one type of reaction centre, discussed in this chapter. The reaction centre of PSII of green plants has some relationship with that of the purple bacteria, and is described in Chapter 5. The reaction centres of PSI (Chapter 6) and of green sulphur bacteria (Chapter 7) may have some affinities with each other, but are quite different from the purple bacterial type.

4.2 Operation of the Reaction Centre

The cell membrane of purple bacteria carries the photosynthetic apparatus, and it is often proliferated into tubes that fill the cell (see Chapter 1). The extent of the membrane invagination, in for example *Rb.* (formerly *Rps*) *sphaeroides*, depends inversely on the light level. After growth in low light, the cells yield on breakage a high proportion of sealed vesicles (chromatophores) that are an intrinsically simple and informative research material for the study of the association between electron transport, PMF and phosphorylation. After high-light growth, the yield is much less (66). *Rps viridis*, on the other hand, develops a folded-sheet structure which does not lend itself to good chromatophore preparations.

The light-harvesting and reaction-centre complexes can be separated and isolated from the membrane material by means of detergent solubilisation and protein fractionation (see Table 4.1).

4.2.1 Biochemistry

Reaction-centre preparations have been obtained from most species of purple bacteria, and there is considerable similarity between them, and between them and the reaction centre of the green bacterium *Chloroflexus* (see Chapter 2). There are three polypeptides, 'heavy' (H) (absent from *Chloroflexus*), 'medium' (M) and 'light' (L). The L and M peptides carry the pigments, four molecules of bacteriochlorophyll (BChl a or b), two of bacteriophaeophytin (BPh a or b), one carotenoid and usually one quinone (see Table 4.2). Apart from small variations in the size of the above peptides, there are two practical differences. The first is on the acceptor side, and concerns the ocurrence of UQ, MK or both. The second is on the donor side, and concerns whether or not there is a c-type cytochrome molecule that remains bound to the HML peptides during the isolation procedure. In *Rb. sphaeroides*, *Rb. capsulatum* and *R. rubrum* there is no bound cytochrome, and the soluble cytochrome c has direct access to the reaction centre. In *Chromatium vinosum* and *Rps viridis* there are tightly bound, hydrophobic c-cytochromes in the reaction centre, and in *Rps viridis* (and *Chloroflexus aurantiacus*) there is a bound c-cytochrome that can, how-

Table 4.1 Isolation of antenna and reaction-centre complexes from purple bacteria.

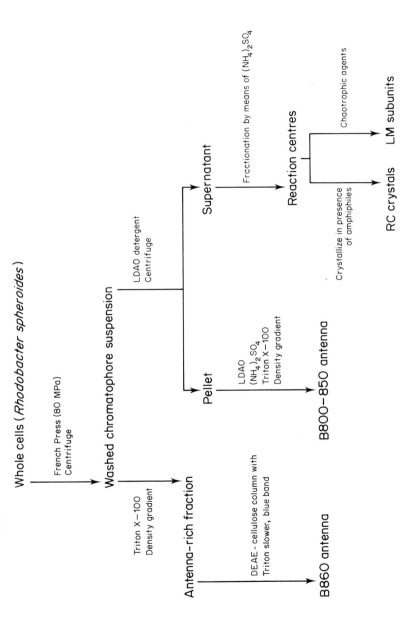

Example taken from P. Haworth, PhD Thesis, University of Manchester, 1979 (adapted). Methods vary between laboratories and for different bacteria. Characteristic spectra are given, for example, by Thornber (14).

Table 4.2 Characteristics of some reaction-centre preparations from purple bacteria.

	Protein (kDa)				Pigments					
	H	L	M	cyt	Bchl	Bph	car	Q	Fe
Rhodobacter spheroides	28	32	36	—	4a	2a	1	UQ	1
Rps. viridis	24	28	35	38	4b	2b	1	MK	1
Chloroflexus aurantiacus	28	30	—	—	3a	3a	—	MK	1

Data from J.P. Thornber (14) (adapted).
There is in each case a second quinone binding site, usually empty when the complex is isolated, which is presumed to carry UQ in the first two cases and MK in the third.

ever, dissociate. The bound c-cytochromes (c-553,c-555) are distinct from the soluble, periplasmic cytochromes c or c_2, and they contribute more than one haem group in any one reaction centre. These are not identical in that there are high- and low-potential forms, sometimes two of each. They are all oxidisable by P870$^+$, and in the case of the high-potential haem oxidation adds to the transmembrane electric field as indicated by the carotenoid shift. This is consistent with their being on the outside (E side) of the membrane with respect to P870. However, oxidation of the low-potential haem has the opposite effect, so logically the low-potential haems should be on the P side with respect to P870, close to the position of Q_A. This is paradoxical (66) and unfortunately occurs in the wrong species: *Rps viridis* has bound cytochrome but no carotenoid shift, and *Chr. vinosum* does not have a crystallographic solution for its reaction centre. *Rps viridis* has a linear array of c-haems, all on the E side of the membrane. Michel and Diesenhofer (68) suggested that they follow the sequence

$$c \ (0 \ \text{mV})\text{---}c \ (0 \ \text{mV})\text{---}c \ (350 \ \text{mV})\text{---}c \ (350 \ \text{mV})\text{---}P960 \ (450 \ \text{mV})$$

This does not explain many features of the oxidation system, such as the faster and low-temperature competent oxidation of the low-potential haems, and it may serve as a reminder that mechanistic problems are not solved at a stroke by the elucidation of structures!

4.2.2 Spectroscopic studies

Figure 4.1 shows absorption spectra of two different reaction-centre preparations, containing BChl a and b respectively. When they are illuminated, or oxidised with potassium ferricyanide ($K_3Fe(CN)_6$), the absorption maxima at 865 nm in (a) and 960 nm in (b) virtually disappear, and are restored by reducing agents such as ascorbate. This matches the behaviour of P700, the photoactive Chl a previously found in chloroplasts by B. Kok, and the same explanation applies here: that one (or two) of the longest-absorbing bacteriochlorophyll molecules have been oxidised, and therefore bleached by the light. There are associated changes in the other peaks, showing interactions between the pigments. The particular bacteriochlorophyll fractions that are bleached by light are labelled P870 and P960 respectively, P standing for photoactive pigment. When flashes of light are used, the spectral bleaching is observed to happen very quickly (picoseconds), and therefore the oxidation is the loss of an electron only. This is confirmed by the persistence of the reaction at liquid helium temperatures: P870 and P960 behave as indicated by the symbol C in the general reaction-centre sequence in Section 3.1.

P870 and P960 are each 'special pairs' of BChl a or b respectively that are close together (see Fig. 4.2); this arrangement was originally deduced from

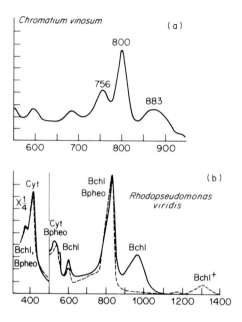

Figure 4.1. Absorption spectra of reaction centres from purple bacteria. (a) A preparation from *Chromatium vinosum* (containing BChl a). (b) A preparation from
Rhodopseudomonas viridis (BChl b). In each case the photo-oxidised spectrum is shown
as a dotted line, showing the bleaching of the active bacteriochlorophyll by the formation
of P880$^+$, P960$^+$. In (b) the absorption of P960$^+$ is shown at 1300 nm. The reaction
centres contain two bacteriochlorophyll molecules forming the P centre, two other bacteriochlorophylls acting as both accessory pigments and as an electron transport path,
and two bacteriophaeophytin molecules, one of which is the acceptor X_0 (see Fig. 3.1).
The three pairs of pigment molecules (see Fig. 4.2) can be seen as discrete spectral
peaks at 756, 800 and 883 nm in (a), but the accessory bacteriochlorophyll and bacteriophaeophytin are superimposed in (b). Photo-oxidation of P can be seen to cause an
electrochromic shift but not a bleaching in the adjacent pigments. After Thornber (69)
and Thornber et al (32)

ESR observations, in which the spectral line due to the unpaired electron was
narrower than expected (by the square root of two). Additionally evidence from
ENDOR (electron nuclear double resonance) showed splitting constants half
those to be expected from the radical cation BChl$^+$. However, the interpretation
of magnetic resonance data from such complex systems is controversial, and the
reader is referred to the review of Norris and van Brakel (71).

   The spectral changes that take place during the bleaching process have been
studied by means of ultra-fast, time-resolved spectroscopy. Laser flashes are
applied both to excite the sample and to measure the absorption spectrum at
a predetermined interval after the flash. The initial changes take place in less

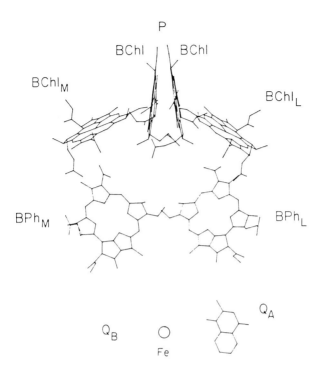

Figure 4.2. The pigments of the reaction centre of *Rps viridis*. Pigment position obtained by X-ray crystallography of reaction-centre preparation, ignoring the protein components (shown in Fig. 4.4). Q_A is MK in this species, and the site Q_B is empty but would bind UQ. Only the pigments on the L-side (adjacent to the L protein subunit) are believed to be active in electron transfer. Reproduced by permission of Dr M.R. Wasiliewsky and Elsevier Science Publishers. From Wasiliewsky, M.R. & Tiede, D.M., (1986) *FEBS Lett.* **204**: 369

time than the duration of the flash, and the experimentally recorded spectra have to be subjected to a computer-based process known as deconvolution. Provided it can be assumed that not more than three or four components exist, a computer calculates three or four consistent spectra that appear and disappear with characteristic times, and which at any instant add up to the spectrum observed.

The earliest spectral form that is consistently observed is that known as P^F or P^+I^- which shows among other effects a loss of absorbance at 540 nm and 760 nm which has been identified as the reduction of BPh a. This P^F state appears in 4 ps and was interpreted as $[(BChl)_2{}^{\bullet+}.BPh^{\bullet-}]$ formed by the expulsion of an electron, passing either through or past another bacteriochlorophyll molecule to bacteriophaeophytin. There is therefore a photo-induced radical pair formed. Parson and Holten (72) have, however, warned that the above ion-pair description is an indication of the sequence of events, but not a proper description in that the optical phenomena may involve complex electronic states of several molecules.

The oxidation of P870 to P^+ can be carried out in the dark by the use of redox buffers, for example ferricyanide–ferrocyanide mixtures of known redox potential. It was found that the P/P^+ couple had a standard potential of +450 mV. The energy of the first excited singlet state of BChl a (960 nm) is 1.35 eV, and the standard potential for the couple P^-/P is therefore approximately −900 mV.

If the electron is unable to proceed further, the state P^F lasts for the unexpectedly long time of 15 ns, and the stabilisation is ascribed to the formation of a multimolecular triplet state that includes the 'bridging molecule' of the intermediate bacteriochlorophyll.

The reaction-centre preparations contain two binding sites for UQ or MK, one of which is usually empty. There is also one Fe^{2+}. It was assumed that the two quinones formed part of the electron transport chain, because of the broad ESR spectrum that succeeded the sharper one of P^F. The role of iron was discounted because it could be replaced by manganese, or removed altogether, with no adverse effects, and furthermore the valency of the iron did not change during the reduction of the presumed acceptor. Loach and Hall showed that when the iron was removed the ESR spectrum became sharper and resembled the known spectrum of the UQ semiquinone (UQ^-), and Cogdell and others were able to reversibly inactivate the reaction centre by extracting and replacing UQ. The original ESR spectra could then be reinterpreted as being due to UQ^- complexed with iron. Hence the second state of the reaction-centre complex was the radical pair $[(BChl)_2^+.UQ^-Fe]$ or more simply P^+/Q^-. The tightly binding site is known as Q_A, and the exchangeable site (empty when prepared) as the Q_B site.

Ultra-fast spectroscopy detected the formation of the free-radical semiquinone anion of Q_A in a period of 150 ps. The ion-pair P^+/Q_A^- persists for up to 0.1 s if the electron cannot progress further. The Q_a is on the P side of the membrane while P870 is on the E side; the formation of the P^+/Q^- pair is electrogenic, that is, it is accompanied by a membrane potential, which can be detected by a shift in the absorption of carotenoids (see Section 4.3). This is known as the fast electrogenic effect, as opposed to the slow one that indicates the electron transport within the bc complex.

Q_A^- does not acquire a proton, nor is it ever reduced further (such as to Q^{2-}). This is presumably indicative of the restrictive nature of the protein environment. Q_A^- is reoxidised by transfer of its electron to Q_B. The Fe^{2+} ion does not appear to be necessary, and presumably only has a structural role. The transfer takes 0.1 ms. Q_B^- waits for a second electron from the next turnover of Q_A^-, whereupon it gains two protons from the aqueous E phase, and becomes mobile in the lipid phase of the membrane. This process has been described as a two-electron gate. The redox potential for the formation of the semiquinone Q^- in the free state is very negative and would constitute a barrier to electron transport. The attachment of Q_A to the protein raises the redox potential, overcoming the problem.

In *Rb. sphaeroides* both Q_B and Q_b sites are occupied by UQ; in *Rps viridis* Q_A is MK, and Q_B is UQ, and in *Chloroflexus* both sites are occupied by MK.

The re-reduction of $BChl_2^{\cdot+}$ to $BChl_2$ coincides with the oxidation of the *c*-cytochrome. Figure 4.3 represents the electron transfers in the reaction-centre complex of purple bacteria, using a redox potential scale as an indication of the energy that is expended following the absorption of light.

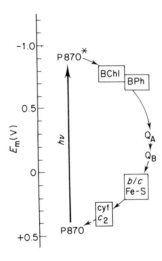

Figure 4.3.   The cyclic electron transport system of purple bacteria arranged on a redox-potential diagram. Excited bacteriochlorophyll (P870*) reacts as the couple P^*/P^+; the electron acceptor is the couple P^+/P. The potential difference between the couples is the energy of the quantum of 870 nm light

4.2.3 The crystallographic solution of bacterial reaction-centre complexes

X-ray crystallography of reaction centres has been achieved with *Rps viridis* and *Rb. sphaeroides*. Figure 4.2 shows the model produced by Deisenhofer et al (73) for the *Rps viridis* complex, just showing the prosthetic groups. They point out the two-fold rotational symmetry, as in a two-pronged fork. The special pair of BChl b molecules is confirmed, and it can be seen that they are mutually in contact via the magnesium atom of one and the acetyl group of the other. The average distance between the plane of the tetrapyrrole rings (which are inclined at $15°$) is 0.3 nm, and the Mg–Mg distance is 0.7 nm. A water molecule between the rings, expected from previous work, is not found. The nearest pigment neighbours to the special pair are BChl b molecules in contact on each side, and these are themselves in contact with BPh b molecules. However, the symmetry is broken in that there is a MK molecule on one side only, which must represent the Q_A site. The iron atom is located centrally and connects the MK molecule to an empty pocket on the other peptide, which is presumed to be the exchangeable Q_B site. The differentiation of the Q-sites implies that there is only one working electron transport route, down the side leading to Q_A. The two BPh molecules have different wavelengths, and it is the longer one which is active. The BChl b and BPh b molecules on the 'dead' side are therefore antenna pigments only.

A nomenclature coming into use is to add the protein subunit as a suffix to the pigment abbreviation. P is used for the special pair, which belongs to both L and M subunits. Hence the electron pathway starts from BChl$_P$ and proceeds via the accessory BChl$_{AL}$ to BPh$_L$. The pigments on the dead side are BChl$_M$ and BPh$_M$.

The four c-cytochrome haem groups are seen to form a conductive chain leading in from the periplasmic space.

Figure 4.4 is a diagram of the associated protein subunits, intended to serve as a key to the computer-generated model of Deisenhofer et al (73). The original was presented as a stereoscopic pair, and the reader is strongly urged to consult it with a viewer. A corresponding set for *Rb. sphaeroides* has been published by Allan et al (243).

4.3 Cyclic Electron Transport

The action of the reaction centre of purple bacteria is to take an electron from reduced cytochrome c and pass it to UQ via the Q_B site, using the energy of light to overcome the unfavourable difference in redox potential. Much of the phototrophy of the purple bacteria can be accounted for by supposing that the electron travels in a closed cycle (cyclic electron transport), returning to cytochrome c. This requires the participation of the particle known as the bc complex, which acts as a ubiquinol:cytochrome c reductase.

4.4 Generation of PMF

Bacterial photophosphorylation is inhibited by the same inhibitors as are mitochondria (such as oligomycin) and uncoupled by the same agents (e.g. valinomycin and potassium ions). It is therefore based on the same chemiosmotic principle and depends on a PMF made up of a membrane potential and a pH difference. The PMF results from the protons translocated (i) by the simple

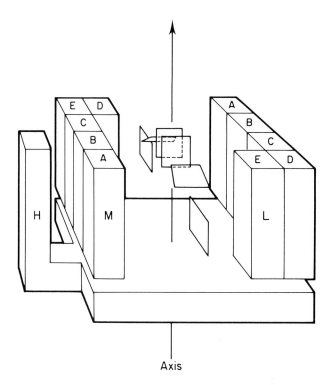

Figure 4.4.   Relation of polypeptides to pigments in the reaction centre of *Rps viridis*. The structure shown is located in the membrane, so that the vertical blocks are the membrane-spanning α-helices of the L, M and H subunits. The axis is normal to the plane of the membrane and points to the E (periplasmic) side. The pigment molecules represented by squares in the centre correspond to Fig. 4.2, and the L and M subunits have been moved apart for the sake of clarity. The sections of the L and M subunits that join the helices A–E are not shown. The original computer-generated stereo drawings, from which this impression has been drawn, also show the protein and the four haem groups of cytochrome c_2 on the E side of the complex. The reader is urged to examine the originals (73) with a stereo viewer

quinone/quinol loop and (ii) by the Q-cycle or similar systems (described in Section 3.3).

Measurements can be made of membrane potentials in various ways. In the case of mitochondria, it is possible if difficult to insert a probe directly. For small vesicles such as chromatophores, perhaps the most general method is the uptake of a diffusible non-metabolisable ion such as tetraphenylborate. The ion will distribute itself so that its concentrations of the two sides of the membrane are in accord with the Nernst equation:

$$E_m = (RT/nF) \ln ([TPB^-]_E / [TPB^-]_P).$$

The concentrations are measured after a quick separation of the chromatophores from their suspending medium. It is not, however, possible to follow the rate of formation of the membrane potential by this method, and a more versatile technique (not applicable to mitochondria) is to observe the electrochromic shift of carotenoids, as follows.

In general, a pigment properly oriented in an electric field will show a change in its absorption spectrum. This is the electrochromic shift, or Stark effect. Although the expected values of the membrane potential are only some 0.1 V, the thinness of the membrane (8 nm) leads to a correspondingly high value of the electric field (12,500,000 V/m). It so happens that the reaction-centre carotenoid responds with a wavelength shift of 1 nm or more, in the 450–550 nm region of the spectrum. If a difference spectrum is recorded, between two cuvettes containing chromatophores in which the only difference is that in one cuvette an artificial membrane potential of known magnitude has been created by means of valinomycin and potassium ions, then it will resemble the first derivative of the carotenoid absorption spectrum with respect to wavelength, and the magnitude is obtained in terms of potential. In general it appears to be a linear relationship.

Some difficulties in reconciling these measurements with the theoretical behaviour expected of such electrochromic effects may be explained by supposing that the particular carotenoid responsible is located in part of the protein which has a pattern of charges that impose a fixed background or bias potential.

The other component of PMF is pH difference. The disappearance of hydrogen ions from the external medium is easy to measure with electrodes or indicators, but there is some difficulty in establishing that the protons are indeed going into the internal E space and not combining with the membrane, as well as uncertainty as to whether the internal E space is buffered. The chromatophores are usually isolated with techniques that could be expected to rupture them, so that when they reseal themselves they would contain proteins from the protoplasmic extract, and hence an uncertain degree of buffering capacity would exist. Nevertheless there is agreement about the pH difference. Probably the most direct method for measuring the internal pH of chromatophores is the use of ^{31}P-NMR, which follows the state of phosphate ions. The results have been shown to validate the use of indicators such as neutral red and acridine orange.

The sequence of events in response to the onset of illumination is that the membrane potential rises to a value of some 50 mV in the time interval expected for the full separation of oxidant and reductant in the reaction-centre complex, but not allowing time for either cytochrome or diffusible UQ to be involved. On this basis the electron simply moves across the membrane thickness from the E to the P side. This is followed by a further, slower rise, to a value of some 0.1 V, accompanied by the development of some 2 units of pH difference.

The slow phase of field generation must be due to electron transport through the *bc* particle, because inhibitors (see Fig. 4.5) eliminate it. It is expected that the reduction of UQ at the Q_B site of the reaction centre by two electrons will take up two protons from the P phase, and these will be released into the E phase when UQH_2 is reoxidised at the *bc* complex. This is the simple quinone–

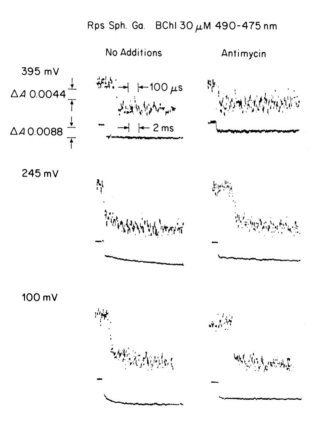

Figure 4.5. Fast and slow electrogenic stages of electron transport. Chromatophores of *Rb. sphaeroides* were adjusted to the redox potentials shown, and their absorption was measured in the carotenoid region during and following a short light flash. The drop in absorbance is due to an electrochromic effect of membrane potential on the carotenoid, and can be seen to have one component which is complete in 100 μs (due to the formation of the pair D_n^+/X_n^- (see Fig. 3.1) in the reaction-centre complex), and a second component which is slower (2–10 ms), and is eliminated by high redox potentials (395 mV) or by antimycin A. The second component is due to electron transport in the cytochrome *bc* complex. Reproduced by permission of Dr J.B. Jackson, Professor P.L. Dutton and Elsevier Science Publishers. From Jackson, J.B. & Dutton, P.L. (1973) *Biochim. Biophys. Acta* **325**: 104

quinol loop and provides 1 H^+/e^-. The cyclic process, such as the Q-cycle described in Section 3.3, makes the value up to 2 H^+/e^-.

4.5 Photophosphorylation

A.W. Frenkel (78) discovered the occurrence of cyclic photophosphorylation in chromatophores of *Rhodospirillum rubrum* in 1954–6, closely following D.I. Arnon's work on chloroplasts.

The broad consensus of the mechanism of phosphorylation by means of the F-ATPase coupling factor has been described in Section 3.4. One important question still remaining concerns the number of protons required to form one molecule of ATP. The original suggestion, for the mitochondrion, made by Mitchell in 1966, was $2H^+/ATP$, but this is not possible to reconcile with the values of the PMF and the observed steady-state concentration quotient for ATP, ADP and P_i. On the basis that whole numbers are to be preferred, the value of $3H^+$ is widely accepted, and more or less established in chloroplast thylakoids, but the actual measurement in bacterial chromatophores is difficult. Values of 2–2.3 were obtained by Petty and Jackson (75, 76), but it appears that the concept of a single cytoplasmic (or even periplasmic) space may be an oversimplification (see the review of Melandri and Venturoli (77)).

Several bacteria are able to use pyrophosphate ($H_4P_2O_7$) as an alternative to ATP, and they may develop solid accumulations of polyphosphate $[(HOPO)_n]$ which probably provides an energy reserve (Fig. 1.4).

4.6 Photoautotrophy: Reverse Electron Transport

The above provides an account of the formation of ATP by cyclic electron transport. However, for autotrophic metabolism a reduced coenzyme needs to be generated. This is believed to come about by the process known as 'reverse electron transport'. In this sense normal electron transport is the respiratory transfer from low-potential systems such as NADH to UQ, and reverse electron transport is the opposite, achieved at the expense of PMF energy in the membrane (see Chapter 3, Section 3.4). In aerobic respiration (as in mitochondria) the difference of 0.4 V in the standard redox potentials of UQ and NAD^+ provides the energy for the pumping of protons to form the PMF and hence ATP. In reverse electron transport, electrons from quinol find their way to NAD^+, drawing the necessary energy from the PMF. The process was reported by Frenkel (78) and established by Keister and Yike (79), in *R. rubrum* chromatophores.

If a coenzyme is to be reduced, an environmental reductant is to be oxidised. Water is not possible because the most oxidised component of the bacterial system is P870 at +0.45 V, and +0.815 V are required to decompose water at pH 7. It is not clear how electrons enter the electron transport pathway from the natural donor but three possibilities exist, depending on the reducing potential of the donor.

Hydrogen gas ($E'_0 = -0.42$ V at pH 7) can reduce NAD^+ directly in the cytoplasm by means of a soluble hydrogen dehydrogenase in some species. Less powerful reductants with potentials of approximately zero (remembering that the actual potential depends on both the standard potential, the concentration and pH, see Section 3.2) may reduce quinone in the membrane by means of a specific reductase complex such as the succinate:ubiquinone oxidoreductase (like mitochondrial complex II). Donors with higher potentials may reduce cytochrome c in the periplasmic space, probably by interaction with specific cytochrome c reductases (the preferred hypothesis), and in this case the electrons would need to pass to the quinone via the photosystem. Figure 4.6 shows the differing requirements of these donors.

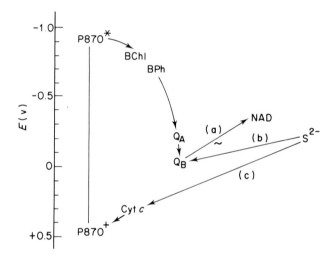

Figure 4.6. Non-cyclic electron transport in purple bacteria. Fig. 4.3, modified to show how NAD^+ can be reduced by energy-dependent reverse electron transport (a), by means of electrons provided by, for example, sulphide (S^{2-}), which could enter the system at the levels of quinone (b) or cytochrome c (c).

4.7 Lithochemotrophy

It may be noted that the means by which inorganic donors reduce components of the electron transport chain are not specific to phototrophic organisms. Either oxygen via cytochrome oxidase or other terminal oxidants may provide the means for supply of ATP and NADH (Fig. 7.7), and hence allow, for example, the reductive pentose cycle to operate. This is 'chemosynthesis', or better 'lithochemoautotrophy'. The bacteria are 'Winogradsky's bacteria', also known as the lithotrophs.

Some purple photosynthetic bacteria are able to grow lithochemoautotroph-ically in the dark using reduced sulphur compounds or hydrogen (H_2), and molecular oxygen, as the energy source.

4.8 Relations between Reaction-centre and Light-harvesting Complexes

Purple bacteria are able to vary the proportions of light-harvesting to reaction-centre complexes in response to changes in the level of light. The photosynthetic unit (BChl per reaction centre) varies from 65 in high light to 126 in low light in *Rb. capsulatum* (80). There is more quinone and *bc* complex in the high-light condition, and more ATPase particles, both in rela-tion to reaction-centre complexes. Drews suggested that when the light level is increased, the electron transport proteins were produced relatively faster and displaced antenna complexes in the membranes, so that the smaller units were more efficient, meaning that the reaction centre turned over more often, al-lowing faster electron transport. The area of membrane is diminished in high light. The mechanism of this control is not understood, although it has been established that the genes for both reaction-centre and light-harvesting (B870) polypeptides are cotranscribed. It is possible that the B800–850 antenna is reg-ulated by the supply of carotenoid.

The size of the photosynthetic unit and its efficiency is regulated by light, probably at the translational level, but the synthesis of the whole photosynthetic apparatus is controlled by oxygen, being inhibited by air and maximally induced by oxygen at a partial pressure of 70 Pa (the partial pressure of oxygen in air is 2000 Pa). The control appears to be exercised via certain promoter sequences in the (RC, B870) operon, and hence is transcriptional.

4.9 A Connection with Respiration

In air, cells of the Rhodospirillaceae carry out respiration, behaving like mitochondria. It is possible that parts of the membrane may be specialised for respiration, although the quinones and the *bc* complexes are common to both. The central role of the *bc* complex is illustrated in Figs. 7.2–7.7.

There is some current discussion (see Ferguson (81)) that the soluble cy-tochrome *c* (c_2, *c*-550) is not necessary for electron transport in *Rb. capsula-tum*, either in mutants where it is absent, or in sphaeroplasts, where it is lost because there is no cell wall to contain it in the periplasmic space. There is some change in the kinetics of the oxidation of cytochrome c_1, and it may be that diffusion of the *bc* and cytochrome oxidase complexes, even though they are massive, may be sufficient to maintain the reaction. The cytochrome c_1 projects from the membrane and may be able to swing sufficiently to make contact with the receptive surface on the reaction centre. There may be a future revision of the function of cytochrome c in bacteria, and, by extension, in mitochondria.

Further Reading

A survey of taxonomy, phylogeny and habit of purple bacteria may be found in:

Pierson, B.K, & Olson, J.M. (1987) Photosynthetic bacteria. In Amesz, J., ed., *Photosynthesis*, New Comparative Biochemistry, vol. 5, Elsevier, Amsterdam, pp. 21–42.

Chapter 5

The Green Plant: Photosystem II

5.1 The Context of PSII in a Two-light-reaction Theory

The significance of PSII, as seen today, is that it is a protein–chlorophyll complex, existing in a specialised membrane, that carries out the light-driven reaction catalysed by water:plastoquinone reductase. The water gives rise to O_2 in the lumen of the thylakoid, and the PQ pool that becomes reduced (to PQH_2) diffuses freely in the lipid matrix of the membrane.

The other light reaction in green plants is driven by PSI (Chapter 6): this is catalysed by plastocyanin:ferredoxin reductase. A single enzyme reduces $NADP^+$ (FNR), and PSII is connected to PSI via the cytochrome b_6f particle, acting as a plastoquinol:plastocyanin reductase.

The existence of two light reactions is general knowledge today, but was a matter of great excitement in the early 1960s, when evidence came together, somewhat haphazardly, resulting in the present theory of the 'Z-scheme' in green plants. Perhaps the early workers (for example Willstatter) were too aware of the similarity of chlorophyll to haem, and looked for an enzyme containing chlorophyll that carried out a photochemical reaction, perhaps on carbon dioxide directly. This tradition clearly motivated O. Warburg, who pursued such a simple account until his death in 1970. However, several lines of reasoning had been developing from considerably earlier. Algal cells provided the material for virtually all these experiments, since at the time chloroplasts were unreliable over an extended period. The usual species was of course *Chlorella*, but other unicellular and laminar algae were also used (see the review by Urbach (82)).

5.1.1 Evidence for two light reactions: the red-drop

If normal photosynthesis in algae is the output of oxygen and the uptake of carbon dioxide according to de Saussure's equation

$$6CO_2 + 6H_2O \rightarrow C_6H_{12}O_6 + 6O_2$$

then the effectiveness of light for normal photosynthesis falls sharply at wavelengths longer than 680 nm, shown by Emerson and Lewis in 1943. That is to say the number of molecules of oxygen produced or carbon dioxide reduced, per quantum of light absorbed, was approximately constant up to 680 nm, but declined to nearly zero by 700 nm. This implied that quanta absorbed by some of the chlorophyll at longer wavelengths than 680 nm were ineffective at operating normal photosynthesis. (See Chapter 2 for discussion of the different physical forms or situations of chlorophyll in the protein complexes, resulting in the absorption band being noticeably broader than that of chlorophyll in, say, acetone solution.)

5.1.2 'Enhancement'

Two alternative beams of light could be set up to illuminate a culture of algae, one beam of wavelength in the region 650–680 nm, and the other in the region 700–720 nm. The intensities were adjusted so that the rates of normal photosynthesis were the same for each light beam, that is, the far-red beam was much more intense. When the two beams were supplied simultaneously, it was observed that the rate was two to three times greater than the sum of the rates obtained with the individual beams separately. This phenomenon became known as the Emerson enhancement effect (83).

5.1.3 Chromatic transients

In a variation of the enhancement experiment, it was shown that if the two light beams were administered alternately, there was a transient burst of oxygen released when the red light followed the far-red one. It was possible to interpose a period of darkness between the turning off of the far-red beam and the turning on of the red. The magnitude of the transient diminished with the length of the dark interval between the two beams. The inference was that there were two photoreactions with slightly different and overlapping absorption spectra, and that one photoreaction was preferentially driven by one beam, generating a product that was a substrate for the other photoreaction (84). The rate of decay of the stored transient increased with increased temperature, and a chemical substance was postulated as an intermediate.

Concurrently, it was shown that the two beams had antagonistic effects. It was known that the yield of fluorescence from a leaf or an algal suspension depended on the history of that preparation, and continued to change while the fluorescence was being observed (the Kautsky effect). The initial yield was shown to be low if the cells had been pre-illuminated with far-red light, and high with red light. It was argued by Duysens that the two light beams had opposite effects on a hypothetical 'quencher' (Q) (85). Far-red light generated Q and lowered the fluorescence yield. (This hypothesis originated with Kautsky (86) on the basis of the change in fluorescence kinetics; it was Duysens who shone the two lights on the problem.) Alternatively, it was possible to measure the oxidation state of cytochrome f using the strong and sharp α-absorption band of the haem at 554 nm, shown by the reduced cytochrome. Far-red light ('light-I') oxidised cytochrome f, and red light ('light-II') reduced it (85). The photoactive chlorophyll P700, discovered by Kok in 1957, was shown to undergo red/far-red changes (87).

5.1.4 Where are the intermediate-potential compounds to be placed?

It was argued by R. Hill and F. Bendall that the existence of cytochrome f (they included cytochrome b_6 although oxidation–reduction changes for it were not established) demanded the existence of two light reactions. To evolve oxygen, the photosynthetic apparatus needs to generate an oxidant with a potential more positive than 0.8 V at pH 7. Also, to reduce $NADP^+$, it needs to generate a reductant with a potential more reducing (more negative than) -0.32 V at pH 7. A single light reaction would take the electron across this gap in one movement, so that materials such as cytochrome f ($E'_0 = 0.36$) could not be fitted onto the direct scheme and could not be part of the direct electron transport chain. They proposed the 'Z-scheme' (88) in which cytochromes took part in a section of electron transport that connected two light reactions.

Either because of its absorption of Duysens' 'light-I', or because it was considered to be more similar to the better-known photosystem of the purple bacteria in being non-oxygenic, the green-plant photosystem that reduced $NADP^+$ was labelled (photo)system I, and the oxygenic one photosystem II.

Continuing from the discussion of redox potentials (above), a one photon/electron scheme would require a photon of 680 nm, which has the energy of 1.8 eV, to expend 1.12 eV in forming the oxidant and reductant above, plus energy corresponding to at least half an ATP molecule per electron, say 0.3 eV, giving a total of 1.42 eV. This makes no allowances for stabilisation energy (the process of photosynthesis is obviously irreversible, so that considerable energy must be lost as heat). 1.42 eV conserved from 1.8 eV was considered to be an unreasonably high efficiency. This argument has gained strength as the details have become better known.

5.1.5 Functional separation of PSI and PSII

The electron transport chain of normal photosynthesis can be divided into partial reactions by the use of artificial electron donors and acceptors, and inhibitors. Oxygen is evolved in the 'Hill reaction' in which an artificial oxidant (electron acceptor) is supplied. The original acceptors used by Hill were ferric complexes such as ferrioxalate (89), which happen to accept electrons preferentially from PSI in unfractionated thylakoids; ambiguities can be diminished by the use of lipid-soluble acceptors such as derivatives of benzoquinone, maintained in the oxidised state by potassium ferricyanide ($K_3Fe(CN)_6$). These are specific for PSII. The Hill reaction, defined as the production of oxygen in the presence of an artificial electron acceptor, shows the red-drop, and is inhibited by compounds such as diuron.

Chloroplasts can be made to photoreduce $NADP^+$ without the production of oxygen if electrons are supplied from ascorbate via a trace of a dye such as DCPIP. Ferredoxin and $NADP^+$ are lost from broken chloroplasts and need to be added, whereupon the reduction of the coenzyme can be followed spectrophotometrically at 340 nm. Alternatively, a viologen dye (such as paraquat) can be photoreduced directly, and used as the agent for the Mehler reaction; the uptake of oxygen is followed. These photoreductions are not subject to the red-drop, and indeed the quantum requirement is approximately halved at long wavelengths where the chlorophyll of PSII does not absorb.

In 1964 Boardman and Anderson fractionated chloroplast thylakoids by means of the surface-active agent digitonin. They obtained small particles which accounted for most of the PSI activity (ascorbate to $NADP^+$), and a heavy precipitate that accounted for PSII (the Hill reaction with potassium ferricyanide); there was little total loss, and the electron transport chain could be reconstituted provided plastocyanin was re-supplied. Mechanical disintegration followed by density-gradient centrifugation was later found to achieve a similar resolution, and it has been shown that the grana of granate chloroplasts contain virtually exclusively PSII.

5.2 The Water-splitting System is the Donor to PSII

Photosystem II has a reputation of being more labile than PSI. This is mainly due to fragility of the oxygen-evolving complex. Heating to 50°C, or incubation with 0.8 M tris, removes or disorders the essential manganese atoms. Mechanical treatment which breaks the thylakoid membrane tends to cause some loss of extrinsic polypeptides that protect the oxygen site on the E side of the membrane. It happens also that an intrinsic peptide (the B-protein or D1) provides a site for attachment of powerful inhibitors (herbicides) such as diuron; there is no analogy for this in PSI.

Assay and measurement of PSII is conveniently carried out by means of the Hill reaction, when an oxygen electrode may be used to obtain a continuous record. If the oxygen-evolving complex is lost, then artificial electron donors must be used, such as hydroxylamine or semicarbazide, and then the reaction can be followed spectrophotometrically, by observing the reduction of the electron acceptor, DCPIP (blue when oxidised) or potassium ferricyanide. These acceptors are subject to inhibition by diuron; recently silicomolybdate has been introduced as an electron acceptor from PSII that is not subject to inhibition by diuron (90) (but see Section 5.5.1).

The production of O_2 is a four-electron process, and no intermediates between H_2O and O_2 can be detected. The existence of a four-electron gate was shown by the experiments in the laboratories of Joliot (91) and of Kok (92), in which it was shown that flashes of light of saturating intensity, and less than 10^{-5} s duration, given to dark-adapted algal cells (and later chloroplasts), produced pulses of oxygen that were unequal and oscillated as shown in Fig. 5.1. The oxygen was detected by a 'rate oxygen electrode' in which the platinum cathode is very large and consumes oxygen rapidly; pulses of oxygen are detected as spikes of current. Dark adaptation of several minutes was necessary.

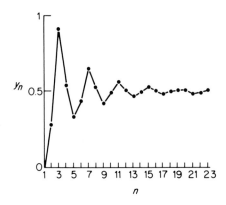

Figure 5.1. Oscillations of O_2 yields of successive flashes. The ordinate, y_n, is the quantity of O_2 released from cells of *Chlorella* by the nth flash in a series. The intensity of the flashes was saturating; the cells had been dark-adapted for 3 min, and the interval between flashes was 300 ms. Reproduced from Joliot et al (91) with permission

The explanation provided by Kok was that a 'store' (S) was accumulating positive charges by successive donations of electrons one at a time to $P680^+$, directly or indirectly. When the store had accumulated four charges, it reacted to produce O_2 and thus returned to its original state. The stages can be represented:

$$2H_2O + S_0 \rightarrow S_1 \rightarrow S_2 \rightarrow S_3 \rightarrow S_4 = S_0 + O_2 + 4H^+$$

(The uptake of water and loss of hydrogen ions could occur anywhere.) The maximal yield on the third flash was interpreted on the basis that S_1 and S_0 were both stable, but during dark adaptation the higher states S_2 and S_3 decayed so that there were three times as much S_1 as S_0 when the flash sequence commenced. The fairly rapid damping of the oscillations was originally explained on the basis that even with apparently saturating light, some 'misses' occurred, or that even with short flashes, some 'double hits' took place. Furthermore, some deactivation could be expected. Discussion of this point is still active.

An alternative explanation was given by Joliot, in terms of cooperation between systems, so that, paraphrasing freely,

$$S_2 + S_2 \rightarrow S_0 + O_2$$

There is little experimental support for connectivity between oxygen-evolving complexes.

5.3 Fluorescence

Photosystem II is the principal source of fluorescence at room temperature. As with any photochemical process, a *yield* can be defined as the light emitted divided by the light absorbed; it is usual to measure light in terms of quanta, particularly if different wavelengths of excitation or emission are used. In the Kautsky effect (93) it was observed that the yield of fluorescence from a dark-adapted leaf rises sharply and declines slowly, and the initial level depends on the past history (see Fig. 5.2).

The fluorescence yield at any instant depends on various factors:

(i) The degree of reduction of the intrinsic electron acceptors (Q) was an early pointer to the two-light-reaction hypothesis (see above). Q stood originally for quencher, but equally validly refers to the plastoquinone Q_A. The rise of fluorescence from the initial point shows inflections which indicate successive pools of acceptors (in fact of PQ), and enabled the abundance of the PQ actually functioning in the thylakoid to be estimated as seven per PSII. (The presence of plastoglobuli means that most of the PQ of the chloroplast is not functioning in electron transport in the thylakoid.) In the presence of diuron, electrons cannot pass from Q_A to Q_B, and the fluorescence rise is therefore quicker; a sigmoidal curve in the presence of diuron indicates that there is a population of connected PSII centres (PSIIα), and an exponential curve (obtained from stroma thylakoids) indicates the isolated centres of PSIIβ (see Section 5.6).

(ii) The S-state sequence also affects fluorescence (see above), provided the

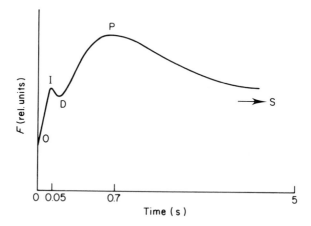

Figure 5.2. Fluorescence induction curve. The ordinate F is the intensity of fluorescence emitted by an alga such as *Porphyridium* when dark-adapted cells are illuminated by steady light beginning at time zero. (Diagrammatic, compiled from several experimental results.) The letters indicate characteristic features which depend on the underlying processes

        flashes are saturating so that the early acceptors in the Q pool are effectively fully reduced. State S_3 (after two flashes) gives the maximum initial fluorescence when the steady light is turned on.

(iii) The electric field across the thylakoid membrane that is part of the PMF develops within a few seconds, and causes the curve to decline from the peak P to the steady state S. In leaves, there are inflections which indicate the onset of carbon dioxide fixation consuming ATP and affecting the PMF.

(iv) The antenna complex LHCII, which by undergoing reversible phosphorylation alters its association with PSII, as well as the stacking of grana in chloroplasts, is believed to cause the state I–state II transitions, accompanied by small deviations in the level of fluorescence about S.

    The fluorescence emission spectrum of chloroplasts at room temperature shows merged peaks at 680 and 695 nm, interpreted as derived from an internal antenna (C670–F680) of PSII and either the PSII reaction centre (C685–F695) or PSI antenna. At cryogenic temperatures these peaks are resolved, and a major peak at 720 appears, due to the reaction centre of PSI (see Section 2.4).

5.3.1 Delayed light emission

    A related phenomenon is delayed fluorescence, or delayed light (DL) emission, discovered by Strehler and Arnold (94) in an attempt to measure photophosphorylation with luciferase; the persistent 'afterglow' vitiated the results. Delayed light has the emission spectrum of normal ('prompt') fluorescence, but whereas prompt fluorescence even of uncoupled chlorophyll has a lifetime of only 2–5 ns (Section 2.4), DL emission has been measured at times ranging from 25 ns to 90 s or more. The intensity of DL emission is inversely proportional, very roughly, to the time interval between the end of the excitation and the observation, for delays between 1 μs and 1 ms. In that time interval the intensity declines from $1/168$ to $1/2900$ and the absolute yield from 4×10^{-6} to 4×10^{-7} (see the compilation of data in the review by S. Malkin (95)), so that DL emission does not represent a significant diversion of energy from photosynthesis. However, the form of the decay curve of the DL emission is such that roughly equal numbers of quanta are emitted in each order of magnitude of time interval; thus the proportions in the intervals 1–10 ms and 1–10 s are 10.8 and 44.6 respectively. This is of interest because it suggests that some chemical or thermal process is re-exciting the chlorophyll from a succession of energy stores, which are in relatively simple molar ratios. DL emission is therefore an indicator of the movement of energy from the active centre along the early stages of electron transport, and arises from a back-reaction in which electrons or positive charges (holes) at these stages are dislodged and return to P680, re-creating P^* which emits the light. There is loss of energy with each electron transfer, and therefore DL emission needs thermal energy for the back-reaction. DL emission is indeed temperature-dependent at times longer than 1 ms.

Delayed light emission in green plants is associated with PSII because of the similarity with prompt fluorescence, its virtual absence from PSI particles and from PSII-deficient mutants, and the relation to oxygen production, below. It has to be presumed that the DL emission comes from the same chlorophyll as prompt fluorescence via P680* (the excited state of the photoactive pigment). P680* is presumably regenerated from the primary charge-separated complex $[P^+.Ph^-]$, by return of the positive charge from some carrier on the oxidising side, and of the electron from some acceptor on the reducing side.

Delayed light emission oscillates in level if saturating flashes are given to a dark-adapted sample, in the same way as oxygen yields. This suggests that DL emission is related to the S-states in the production of oxygen. From the fact that the maximum DL emission, like oxygen yield, occurs on the 3rd (7th, 11th, etc.) flash, for 1 ms DL, S_4 is implicated, that is, the four-charged intermediate that immediately reacts to give oxygen. At longer times the maximum occurs on the 2nd (6th, 10th, etc.) flash, implicating S_3. The ADRY reagents (96) which include the uncoupler CCCP, are known to deactivate S_3 and S_4, and are observed to abolish DL.

At longer times, a two-fold oscillation can be detected in DL emission, which is indicative of the involvement of the two-electron gate at Q_B. It is therefore clear that the rate-limiting step in the DL reaction may be either the return of the positive charge or the electron to the primary charge-separated complex.

5.3.2 Thermoluminescence

A related phenomenon is thermoluminescence, or 'glow curves'. A sample is frozen down to 77 K under illumination, and then darkened. All the possible energy stores are presumed to be filled. The sample is then warmed at a known rate, and bursts of light are observed to be emitted as certain temperatures are reached. At a first approximation, the thermodynamic product RT (gas constant $\times$ absolute temperature) indicates the activation energy required to dislodge electrons and reactivate chlorophyll. Luminescence has also been observed from pre-illuminated material by subjecting it variously to a 22–56° temperature jump, addition of solvents, an a.c. electric field, pH changes and addition of salt. Common links between the last four are the transmembrane electric field and pH gradient that contribute to the PMF. Work in the laboratories of Crofts, Fleischman and Barber established that the PMF was a direct determinant of DL emission at 1 ms, that the value of the PMF was approximately 100 mV in steady illumination, and that the activation energy needed for 1 ms DL emission was 0.6 eV. The energy of the emitted quantum at 680 nm was 1.8 eV, so the energy level of the electron pool from which DL emission was obtained was 1.2 V.

Delayed light emission, and thermoluminescence, are affected oppositely to fluorescence by the PMF: DL emission is increased, but fluorescence is diminished.

5.4 The Core Complex of PSII

Photosystem II is a large complex, containing the peptides shown in Fig. 5.3. Unlike the LHCII and PSI complexes, PSII is labile in electrophoretic systems employing sodium dodecylsulphate detergents. With milder systems such as octylglucoside, dodecylmaltoside, deoxycholate or lithium dodecyl sulphate in the cold, PSII particles can be isolated, often with some photochemical activity remaining (see Section 5.8). This has allowed the identification of the constituent polypeptide–chlorophyll complexes. The two largest, known as CP47 and CP43 from their apparent molecular masses on polyacrylamide gel, contain most of the chlorophyll, and Green and Camm (36, 97) proposed that CP47 was the seat of the reaction centre (P680) and that CP43 was an accessory. The genes for the proteins have been sequenced, and the true expected masses of the two peptides, excluding chlorophyll, are 51 and 44 kDa. Diner and Wollman (98) estimated that there were some 40 Chl a per P680 centre in PSII complexes, and most of it is attached, more or less equally, to CP47 and CP43. A third chlorophyll–protein complex, CP29 (22) contains some Chl b, but less proportionately than LHCII (it appears to have little sequence homology with LHCII, but antibodies to CP29 cross-react with LHCII (28)). It is considered to be an interior antenna or linking complex between the site of attachment of LHCII and PSII.

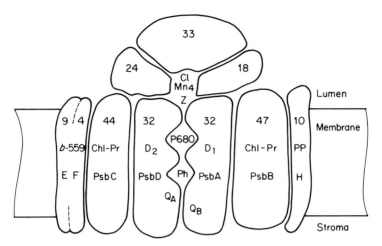

Figure 5.3 A recent model of the PSII unit. The diagram shows the major intrinsic and extrinsic membrane polypeptides that are known to exist in the PSII unit, identified by their apparent molecular mass and gene label (*Psb-*). The novel features of this arrangement are that the D_1 and D_2 peptides carry the electron transport system from the donor Z via P680 and phaeophytin to the $Q_A FeQ_B$ complex, as opposed to the 47 kDa polypeptide. Shapes arbitrary. PP = phosphoprotein

Other peptides may be expected to carry chlorophyll. These are the A- and B-proteins, also known as D_1 and D_2, of apparent masses 32 and 34 kDa respectively. It is agreed that D_1 carries the Q_B site where PQ binds to be reduced by electron transfer from PQ at the Q_A site, presumably on D_2. D_1 is the site of binding of the herbicides diuron and atrazine, which displace PQ. They do not bind exactly at the Q_A site, however, since removal of the projecting tail of the D_1 protein by trypsin protects the thylakoids from inhibition by these herbicides. The two proteins are coded by chloroplast genes *Psb*A and *Psb*D respectively, which have considerable homology; the gene *Psb*D overlaps the sequence of *Psb*C which codes for CP43.

Many of the polypeptides of photosystem II have been associated with gene-products coded in the chloroplast or nucleus (see Table 13.2).

The excitement concerning the D_1 and D_2 proteins has arisen with the demonstration that the genes have significant homologies with those for the L and M subunits of the purple bacterial reaction centre (33, 97). These subunits bind quinones (see Section 4.3.3.2) and are connected by a bridging iron atom according to the X-ray crystallographic model of Michel and Deisenhofer (34, 68), between the four histidines M-217, M-264, L-190 and L-230 (Fig. 5.4). P960 is held between the two histidines M-200 and L-173. It would be possible to adapt the model to PSII by substituting D_2 for M and D_1 for L. The six histidines, in

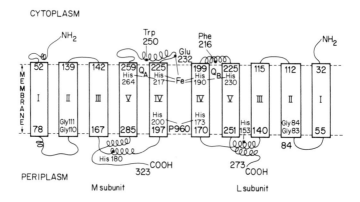

Figure 5.4. Model of the L and M subunits of the *Rps viridis* reaction centre. Diagram based on the X-ray crystallographic model (73) showing the α-helices spanning the membrane between the cytoplasmic and periplasmic faces, connected by links. The attachments of P960, Q_A, Q_B and Fe are shown with some amino acids for comparison with Fig. 5.5. Reproduced by permission of Professor J. Barber and Elsevier Publications Cambridge. From Barber, J. (1987) *Trends Biochem. Sci.* **12**: 321–326

the same order, would be D_2-215, D_2-269, D_1-215, D_1-272, D_2-198 and D_1-198 (Fig. 5.5). (See the review by Barber (37).)

Work on PSII has been greatly hampered by the excessive quantity of chlorophyll compared to the few bacteriochlorophyll molecules in the reaction centres of purple bacteria. The D_1D_2 hypothesis offers hope of spectroscopic advances in the mechanism of PSII—if it is true. There is a recent preparation of a chlorophyll-binding complex obtained by means of Triton X-100, which contained D_1,D_2, cytochrome *b*-559 (see below), five Chl a, two Ph a, a β-carotene and an iron atom. Unfortunately the activity of the complex was limited, one reason being that there was no quinone present. Nevertheless the identity of P680 as a special pair of Chl a molecules has been advanced, and the promise of improved spectroscopy substantiated.

The B- or D_1-protein is also the site of photoinhibition, which is the severe loss of photosynthetic activity that results when plants are exposed to more than saturating light levels (see the review by Kyle & Ohad (99)). This may be due to a direct side-reaction of the B-quinone with the protein, or damage caused by an oxygen species such as superoxide, produced by a side reaction. In some way the damage exposes the protein to the action of a protease that removes it. Replacement molecules are continually being synthesised (the mRNA for this protein is the most abundant in higher-plant chloroplasts). Photoinhibition is a separate topic from the irreversible oxidative damage to lipids and pigments that follows the loss of electron transport from any cause.

The 9 (or 10) kDa phosphoprotein (see Section 2.2.2), judging from its homologies with (part of) the LHCIIβ peptides, may also carry chlorophyll.

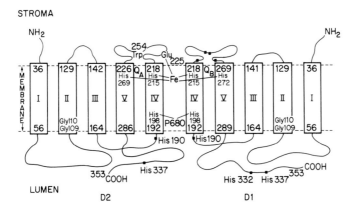

STROMA

LUMEN

D2   D1

Figure 5.5. Model of the D_1 and D_2 subunits of PSII. The polypeptides are arranged so as to stress their similarity with the L and M subunits shown in Fig. 5.4; if these polypeptides carry the electron transport system as suggested in Fig. 5.3, then the analogy with the purple bacterial system provides a hypothesis for their arrangement. cf. Fig. 5.11. Reproduced by permission of Professor J. Barber and Elsevier Publications Cambridge. From Barber, J. (1987) *Trends Biochem. Sci.* **12**: 321–326

Cytochrome *b*-559 exists as two polypeptides, 9.4 and 4.4 kDa, coded by chloroplast genes *Psb*E and *Psb*F respectively. By analogy with the purple bacteria one might have expected a *c*-type cytochrome. Discovered by Bendall (100) it has a high-potential form, normally 0.37 V at pH 7, and a low-potential form of approximately zero. There appears to be a 70 : 30 mixture of the two in thylakoids, and the high-potential form is reduced while the low-potential form is oxidised. The oxidised state (only) can be detected by ESR. Cytochrome *b*-559 can be both oxidised and reduced by P680. In neither case can its action be related to the functioning of PSII (by the argument of Section 5.1.4). Bendall, in a useful review (101), has concluded that if it has a function it may be a minor path around PSII, perhaps deactivating the high-potential carriers when photosynthesis is unable to proceed. The plane of the porphyrin is transverse to the membrane, which may support that view. Bergström and Franzen (102) have shown that removal of the extrinsic PSII 16 and 24 kDa proteins causes all the cytochrome to take on the low-potential state; re-addition of the proteins and calcium ions restores the high-potential condition, as if by re-establishing a hydrophobic environment.

There is no peptide identification for the oxygen-evolving site yet. The three extrinsic, luminal, nuclear coded 16, 23 and 33 kDa peptides (Fig. 5.3) are probably concerned with stabilising or protecting the centre, with the aid of calcium ions and chloride ions, but the centre itself, containing the manganese cluster, is elsewhere (see Section 5.5.2).

5.5 The Operation of the PSII Reaction Centre

The details of electron transport in PSII that follow are summarized in Fig. 6.10 together with those of PSI.

Fast optical spectroscopy has up to now been less rewarding than with the purple bacteria, because the latter yield a reaction-centre preparation containing only six pigment molecules, of which two are active. PSII has been more difficult because not only is there contamination from the extrinsic LHCII, but also there are some 40 chlorophylls in the intrinsic antenna so that the P680 chlorophyll is only a small minority. Nevertheless, PSII possesses some groups or compounds that are unique in thylakoids, and non-optical spectroscopy directed at these entities is proving very rewarding, particularly on the donor side of the reaction centre. For example, the resonance techniques of ESR and ENDOR have detected unusual signals with novel interpretations (Z, see below), and ^{35}Cl-NMR has successfully followed the binding of the essential element chlorine to the proteins protecting the oxygen-evolving site. The other essential inorganic material is manganese, which does not generate a useful resonance signal, but has been studied by means of the X-ray absorption at the manganese K-edge, and by the technique EXAFS (extended X-ray absorption fine structure), confirming the likely existence of an Mn_4 cluster with oxygen bridges.

5.5.1 The acceptor side of PSII

P680, the photoreactive Chl a in PSII, was discovered in Witt's laboratory (103) by repetitive-flash spectroscopy; it is referred to as Chl-a$_{II}$ by this group (see Fig. 5.6). The state of the molecule responsible has been debated, with the conclusion (104) that the ground and first singlet excited states consist of two strongly interacting Chl a molecules, more or less coplanar with the membrane; when P680$^+$ is formed, one ground-state chlorophyll appears. P680 is therefore probably dimeric, but P680$^+$ monomeric. This has been supported by studies on Satoh's D1-D2 preparation.

Phaeophytin a was detected by Klimov and Krasnowskii (105), from the difference spectrum obtained when chloroplasts were illuminated under conditions when Q_A was chemically reduced (i.e. at redox potentials below -0.3 V). The standard potential of the intermediate was -610 mV. Optical and ESR spectra were consistent with the formation of a radical anion Ph$^-$. The phaeophytin is oriented at right angles to the plane of the membrane and of P680 (106) (Fig. 5.7); this is shown by means of a technique in which thylakoids are embedded in a gel (with random orientations). The gel is compressed and expands in one dimension, and the thylakoids tend to become more parallel. Linear dichroism, which is the difference between the absorption of plane-polarised light beams with their electric vectors parallel and perpendicular to the expansion plane, reveals those absorption dipoles of chlorophyll and phaeophytin that are differentially oriented.

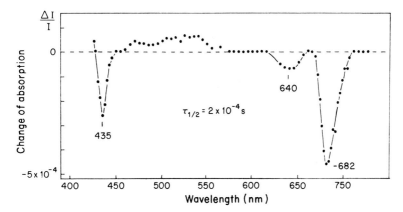

Figure 5.6. First detection of P680. Chloroplasts were enriched in PSII by fractionation with digitonin. The difference spectrum of the Chl a_{II} signal was measured by means of a repetitive-flash technique. The spectrum shows that the loss of absorbance (bleaching) following the flash is of the chlorophyll type. Reproduced by permission of Professor H.T. Witt and Verlag der Zeitschrift der Naturforschung. From Döring, G., Renger, G., Vater, J. & Witt, H.T. (1969) *Z. Naturforsch.* **24b**: 1140

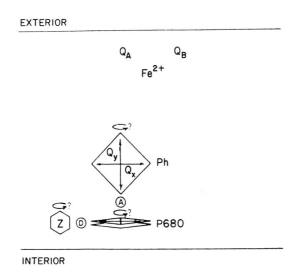

Figure 5.7. Orientation of P680 and phaeophytin in the PSII reaction centre. Measurements of linear dichroism of particles immobilised in gels, when the gels were compressed, provide evidence that the plane of P680 is parallel to the plane of the membrane and at right-angles to the *y*-axis of phaeophytin. Z and D are donors (see Section 5.5.2). Reproduced with permission from Ganago et al (106)

Since P680 is towards the E side of the thylakoid, and Q_A is towards the P side, phaeophytin is well placed to conduct electrons by means of its π-orbital system. In the absence of time-resolved spectroscopy, there is no rise-time value yet for the formation of the pair $[P^+.Ph^-]$; several hundred picoseconds are needed for the electron to reach Q_A.

There is little loss of energy in the formation of the charge-separated pair; taking the energy of the excited state P680* 1.8 eV, the energy of $[P^+.Ph^-]$ was estimated to be 1.75 eV (107, 108). Taking a value of -610 mV for the couple Ph/Ph^-, a potential of about 1.1 V was calculated for the couple $P680/P680^+$. Shuvalov and Klimov (108) suggested that the recombination of the primary charge-separated pair was responsible for most of the prompt fluorescence on the basis of constant properties of the fluorescent emission during the time-course of induction, but time-resolved fluorescence decay measurements, and comparative studies with mutants apparently lacking only P680 (which have very similar fluorescence characteristics to the wild type) (109), militate against the hypothesis. Secondly, although there is little loss of energy in going from P680* to $[P^+.Ph^-]$ (0.05 eV), even that small loss should greatly reduce the PSII fluorescence at very low temperatures, contrary to observation.

The Q_A and Q_B sites

Q_A, the first quencher (Q) shown by Duysens and Sweers (110) to be active when oxidised and inactive when reduced, was identified with a bound quinone (111) in Witt's laboratory when a flash-induced absorbance change was found that matched the state of the quencher. This substance had a light-minus-dark difference spectrum with a minimum at 270 nm and a maximum at about 320 nm and was labelled X-320. The production of X-320, like the inactivation of the quencher, was brought about by PSII but was not inhibited by diuron. At that time the fastest spectroscopic rise-time that could be measured was 20 ns, and both the quencher and X-320 reacted in that time. The decay of X-320 matched the reactivation of the quencher in 600 μs. Bensasson and Land (112) published spectra of the radical anion plastosemiquinone, produced by means of pulse radiolysis in vitro, with the same difference spectrum (PQ^--PQ), which was successfully matched to a more precisely determined spectrum of X-320 by van Gorkom (113) (Fig. 5.8).

The formation of the plastosemiquinone anion induces an electrochromic effect in the spectrum of phaeophytin, resulting in a negative difference-absorption peak at 548 nm, labelled C-550 (114). This effect was found to depend on the presence of β-carotene (115) responding electrochromically to the reduction of Q_A. $Q_A{}^{\cdot-}$ is closely complexed with an iron atom (116), as is the case in purple bacteria.

Redox titrations, using mediators, of the fluorescence yield or the C-550 effect, have indicated that the couple (Q_A/Q_A^-) has $E_{m,7}$ of approximately 0 V, with a dependence on pH of 60 mV/pH. This is surprising since the ESR

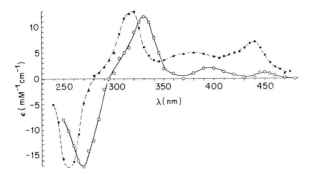

Figure 5.8. Identification of X-320 with plastosemiquinone. The solid line indicates the light-induced spectrum of the primary acceptor of PSII in subchloroplast fragments prepared by means of deoxycholate compared with (dashed line) the plastosemiquinone anion (112). The shift in wavelength was found to be a specific effect of the detergent (113). Reproduced by permission of Dr H.J. van Gorkom and Elsevier Science Publishers. From van Gorkom, H.J. (1974) *Biochim. Biophys. Acta* **347**: 441

data shows no protonation of the A-quinone. The proton uptake in the titration is either too slow to be relevant or in a different site.

Q_A^- reduces Q_B in 600 μs, without apparently making use of the bridging Fe^{2+} as an intermediate. Q_B is the two-electron gate, and takes up two hydrogen ions so as to form PQH_2, which is exchangeable with diffusible PQ in the PQ pool. The Q_B is displaced by herbicides, such as diuron and atrazine, that attack the B (D_1) protein.

There are two apparently trivial observations on the Q_B system that may together become very relevant and practical. As stated above, diuron binds one-to-one with the B-protein and displaces Q_B so that electron transfer from Q_A^- is prevented. Partial digestion with trypsin removes a part of the B-protein, and diuron no longer binds, and has no inhibitory effect. The reagent silicomolybdate, at pH 7, is able to accept electrons from (normal) PSII in the presence of diuron, and probably reacts with Q_A (117); it appears that it has a permanent effect on the B-peptide, making the binding of diuron innocuous. Again, it was known from the early experiments of Warburg that carbon dioxide was essential for photosynthesis, even for the Hill reaction where there was no fixation of carbon dioxide. (This supported his belief in a one-reaction process.) The inhibitory effect of carbon dioxide depletion is manifested at the point of reduction of Q_B (118). The carbon dioxide is tightly bound, probably to the iron (Fe^{2+}) ion between Q_A and Q_B, and may involve hydrogen carbonate (HCO_3^-) ions as the active form; it can be removed by washing with formate solution. Not only do these observations bear on an intimate process in the functioning of the PSII reaction centre, apparently well protected by protein, but they may also lead to technological advances in herbicide development for agricultural use.

5.5.2 The donor side

$P680^+$ is reduced with complex kinetics that depend on the S-states; S_0 and S_1 reduce $P680^+$ in 30 ns, but the time is slowed to 400 ns for S_3. If the oxygen-evolving system is bypassed or damaged, the lifetime of $P680^+$ is increased to microseconds.

$P680^+$ does not have an associated ESR signal, unlike P700, probably because its lifetime is too short for existing techniques. The ESR signal that is associated with PSII is known as signal II, and has several components, II_u, II_s, II_f and II_{vf} (the subscripts indicate the rate of dark decay from unchanging to very fast). Signal II_{vf} is seen in PSII preparations in which the oxygen-evolving system is intact; signal II_f is seen after removal of manganese by tris. With the possible exception of II_{vf}, which is technically difficult, the properties of the ESR line are similar (g=2.0044, dH_{pp}=19 G) and they are thought to be due to the same material, labelled Z (or D, or Y). The rise in signal II_f matches the optically observed decay of $P680^+$, and Z is a candidate for the immediate donor to P680 (see the review by Diner (119)). Optical spectra present difficulty because of the strong but variable electrochromic effects of neighbouring pigments. Redox potentials of Z have been recorded in the range 760 mV (II_s) to 1.0 V (II_f), which are close to the expected value of 1.2 V for $P680/P680^+$. It was suggested that Z might be a cation radical formed from plastoquinone (PQH^+). This surprising concept requires that the protein environment prevents PQH^+ losing a proton, in the same way that the A-protein prevents Q_A from gaining one. ENDOR data (121) support the hypothesis for the ESR signal II_s (regarded as a second radical (D) of the same type as Z). Recent work now points to tyrosine residues instead of PQ as candidates for Z and D (120). The phenol ring system is equally compatible with the ENDOR and ESR results, and in addition two experiments have been performed with

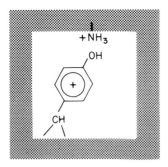

Figure 5.9. Diagram to illustrate a tyrosine residue on the D1 protein in an oxidised form stabilised by its protein environment. Such a structure could account for the ESR signals II_f and II_u, which are associated with the primary electron donor to P680 in photosystem II. Based on an earlier hypothesis (involving plastoquinone) by O'Malley et al (120)

the cyanobacterium *Synechocystis*. First, when deuterated (^{2}H-) tyrosine was supplied to the cells under conditions such that it would be incorporated into protein without degradation and mixing of the isotope, it was observed that the linewidth of signal II became narrower (244). Secondly, using the technique of site-directed mutagenesis, and a transformable strain of the organism (see Chapter 13) tyrosine-160 of the D_2 protein was specifically changed to phenyl-alanine. Signal II_u (interpreted as D^+) was lost, while photosynthetic growth continued slowly (245). The current hypothesis is that tyrosine-161 of the D_1, protein is Z (D_1) and tyrosine-160 of D_2 is the intermediate D(D_2) (Fig. 5.9.).

Manganese, calcium, chloride and oxygen formation

Manganese ions are bound to functional PSII particles, with various degrees of affinity. Treatment with elevated temperature (60° C for 5 min) or 0.8 M tris solution (unbuffered) deplete the manganese content and inhibit oxygen production, somewhat reversibly. None of the individual peptides extracted from PSII particles show much affinity for manganese, and it is probable that the cluster of four Mn is protected but not ligated by the three extrinsic membrane proteins in the lumen of the thylakoid that form a cap over the cluster (122) (see the review by Yocum (123)). Calcium ions are required for the stability of this structure. The three proteins, of masses 33, 24 and 17 kDa, were discovered when broken and re-sealed thylakoid vesicles were prepared from which inside-out and rightside-out populations were separated by partition between non-miscible aqueous solutions of dextran and polyethylene glycol. Inside-out vesicles were readily inactivated by treatment with high salt; the three proteins were found to be detached (124). None of the three have any affinity for manganese, but manganese is rapidly lost if the 33 kDa peptide is removed. Manganese is retained if there is a high concentration of calcium ions.

The requirement for chloride is absolute for oxygen evolution, and the S-clock (p. 103) does not advance past S_2 in its absence. The high affinity depends on the presence of the 23 kDa peptide; in its absence higher concentrations of chloride are needed.

The formation of oxygen takes place on the luminal side of the thylakoid, and is subject to the low pH (approximately 5). This raises the standard potential of the H_2O/O_2 couple from 0.8 V (at pH 7) to 0.9–1.0 V, still allowing the process to be driven by $P680^+$ (1.1 V, see above), but with little wasted energy. One function of the manganese cluster must be to diminish the differences in the energies required to draw successive electrons, because if the first electron is energetically too cheap, the last will be too expensive (preserving the average at 0.9–1.0 V).

The mechanism of the cleavage of water to release oxygen must be related to the S-states, which are confidently expected to be related to the Mn_4 cluster. It has been possible to identify which S-state transitions are associated with the loss of the four H^+. There is general agreement that the pattern is 1-0-1-2, meaning that one H^+ is released when S_1 is formed, none for S_2, one for S_3 and two between S_3 and the release of oxygen and return to S_0. Renger has pointed out that the protein is almost certainly interacting with the Mn_4 cluster, and probably buffering the protons; the 1-0-1-2 cycle observed for the intact system does not necessarily represent that actual sequence at the cluster itself.

At the 1986 Triennial Conference on Photosynthesis, several workers put forward schemes for the involvement of manganese in oxygen production. The cubane-like Mn_4 complex suggested by Brudvig and de Paula (125) and shown in Fig 5.10 is one possibility currently under discussion. This hypothesis relating the S-states to oxy-complexes of manganese, can be associated with a suggestion for the role of chloride, put forward by Coleman and Govindjee (126). The technique of ^{35}Cl-NMR allows the detection of chloride present in PSII preparations, and indicates its atomic neighbours. It was found that there are two populations of chloride, one (intrinsic) associated with the manganese, and one located on the extrinsic 33 kDa peptide. The intrinsic site has three binding sites with pK values in the range 5.4–6.5 which are probably histidine residues. There are eight histidine residues on the C-termini of the D_1 and D_2 peptides that may be considered, plus others found in the loops of the peptides that connect on the E side the membrane-spanning α-helical regions. Their model appears in Fig. 5.11. They go further and propose that chloride and calcium ions interact with the extrinsic proteins, so as to allow them to take up protons from the manganese cluster, sequentially, by progressively rupturing salt-bridges between the protein domains. Their suggestions are based on comparisons of amino acid sequences. While these ideas are obviously speculative, they are nevertheless exciting, and underline the point that once again studies on almost any biochemical process come down to the study of protein.

5.6 Heterogeneity of PSII

There are two populations of PSII, known as PSIIα and PSIIβ. They were postulated by Melis and Homann (127) to explain the kinetics of fluorescence rise at the onset of illumination, in the presence of diuron (see Section 5.3). PSIIα caused a rapid, sigmoidal rise, and PSIIβ a slower, exponential rise (see Fig. 5.12). These two fluorescence components responded differently to magnesium ions (128). It is suggested that the α-population is located in chloroplast grana and the β-population in the stroma lamellae. Although the two are of approximately equal abundance, most work has been done with PSIIα. Time-resolved fluorescence decay measurements have revealed two PSII effects (see Chapter 2), only one of which has a Chl b component in its excitation and was regarded as being due to PSIIα. The larger particles seen on the EFs faces of thylakoids (see Section 1.2) are likely to be PSIIα–LHCII units, and the smaller EFu particles PSIIβ units (without LHCII).

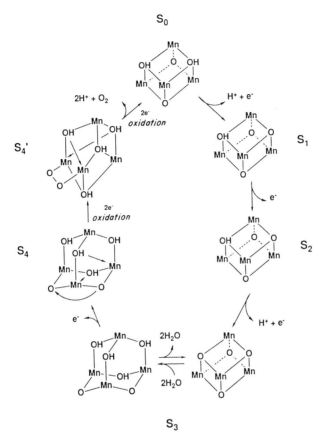

Figure 5.10. A scheme proposed (125) for water oxidation. The O-bridged cubane structure provides a model such that the postulated reactions have reasonable energetics. An important question is the possible binding of protons to protein before their release into the aqueous phase. Note that this scheme is one of several currently put forward. Reprinted by permission of Professor G.W. Brudvig and Kluwer Academic Publishers. Copyright © 1986 by Martinus Nijhoff Publishers. From Brudwig, G.W. & de Paula, J.C. (1987). In Biggins, J., ed., *Progress in Photosynthesis Research*, Proc. VIIth Int. Congr. Photosynth. 1986, vol. 1, p. 497

5.7 The Relation of PSII to LHCII

Although LHCII is a prominent contaminant of PSII preparations, the bulk of the antenna complex does not necessarily remain attached. The separation of PSI and PSII achieved by means of digitonin by Boardman and Anderson (129) depended on a high salt concentration to force the association of LHCII and PSII into a precipitable complex; without the salt the separation does not work.

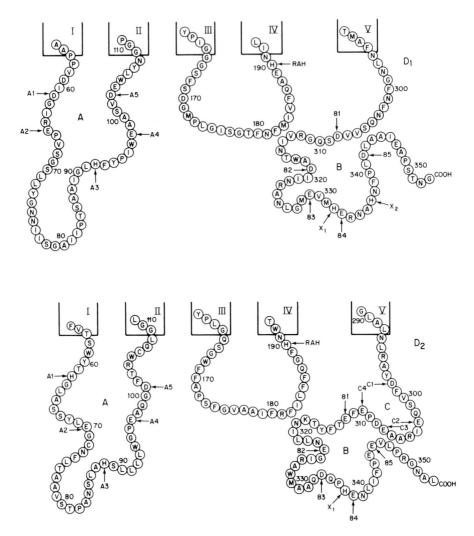

Figure 5.11. Searching for ligands for Mn in PSII. In this study (126) the sequences of the D_1 and D_2 polypeptides were scanned for possible insertions, non-homologous to the bacterial L and M peptides (cf. Figs 5.4, 5.5). The diagram indicates sets of amino acid residues in the non-helical sections of the polypeptides that could provide sites for four Mn ions. In each protein, two manganese-binding sites are proposed, A and B with 5 amino acid ligands for each, and a site on D_2 for Mn/Ca is labelled C_{1-4}. X1, X2 are histidines thought to be important in chloride binding. Amino acids: D = aspartate. E = glutamate. H = histidine. The tyrosines (Y) shown in helix III in D_1 and D_2 are thought to be redox intermediates Z and D respectively, where Z is the immediate electron donor to $P680^+$. Reprinted by permission of Kluwer Academic Publishers. Copyright © 1987 by Martinus Nijhoff Publishers. From Coleman, W.J. & Govindjee (1987) *Photosynthesis Res.* **13**: 206

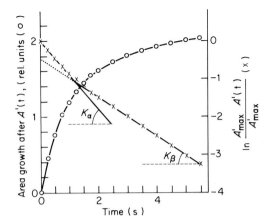

Figure 5.12. Distinction between PSIIα and PSIIβ. In a fluorescence induction experiment, the growth in the area under the induction curve was measured. The kinetic analysis shows two rate constants. Reproduced by permission of Professor P.H. Homann and Pergamon Press. Copyright (1976), Pergamon Journals Ltd. From Melis, A. & Homann, P.M. (1976) *Photochem. Photobiol.* **23**: 346

It is argued that LHCII is not always fixed in its association in the thylakoids in the granum, given that LHCII is phosphorylated by an endogenous kinase in the chloroplast and that this kinase is controlled by the level of reduction of PQ. The hypothesis is that overactivity of PSII reduces PQ and activates the kinase. The phosphorylated LHCII detaches from the PSII, reducing its effective optical cross-section and hence its photochemical activity. A phosphatase of constant activity provides the balance. One problem is that there is no indication of the great increase in fluorescence that could be expected from unattached LHCII (see Chapter 2). It may be that LHCII attaches to PSIIα units in the unappressed thylakoids, or to PSI, but corroborative evidence of Chl b in the action spectra of these complexes is missing.

To some extent the data may be explained by association of the PSII system in the grana into one large 'lake', and dissociation into small arrays ('puddles') in state II. Quanta of light will show 'shot noise', that is, statistical fluctuations that are the more severe the smaller the area under consideration. Even without the dissociation of LHCII from PSII, there will be a much lower quantum efficiency from puddles than from the equivalent lake, because the puddles cannot share out their misses and double hits.

5.8 Preparations of PSII

This section will indicate some of the methods by means of which oxygen-evolving PSII particles have been separated from PSI. It will be apparent that preparation of working, multi-subunit protein complexes from membranes is an exercise in biochemical techniques that has a very wide application in cell biology.

Cyanobacteria have provided a useful starting point, since there is no LHCII, and the main antenna, the phycobilisome. is easily detached. The thermophilic species *Phormidium laminosum* provided active PSII material which performed the Hill reaction at a rate of 1300 μmol O_2 (mg Chl)$^{-1}$ h^{-1} in a LDAO-fractionated, membranous particle (130). The principal difference between cyanobacterial and higher-plant preparations is the frequent presence in the former of a 70–80 kDa linker peptide containing allophycocyanin, and the absence of the extrinsic membrane 17 and 23 kDa proteins. Schatz and Witt (131) applied a related, zwitterionic detergent SB12 to *Synechococcus* sp. and obtained particles with 60–70 Chl a per PSII centre. Extraction of these particles with β-dodecylmaltoside yielded particles with essentially the same chlorophyll ratios, but free from membrane (132); electron microscopy showed spheroidal particles, loosely aggregated in strings, with axes approximately $16 \times 11 \times 6$ nm, and masses of 300 kDa. 300 kDa was expected from the protein and Chl a composition, but a second particle, of 500 kDa mass, was present, interpreted as a dimer. The authors point out that since there are two phycobilisomes per PSII, dimeric PSII might be the natural form. In this preparation the detergent has attached to the particles in undetermined amounts.

Higher-plant chloroplasts contain grana from which PSI is virtually excluded. Grana can be detached from stroma thylakoids by mechanical means (pressure cell and needle-valve), and the two sets of fragments can then be separated by density-gradient centrifugation. Sonic treatment can then be applied, with detergents such as Triton X-100, to break up the grana thylakoids and cause them to re-seal back-to-back, forming small, new vesicles, inside-out with respect to the original thylakoid (133). Obviously this material (known as BBY from the authors' initials) contains both membrane material and LHCII.

A soluble protein complex evolving oxygen and containing all the expected PSII polypeptides has also been isolated from spinach chloroplasts by means of β-dodecylmaltoside, by Franzen (134). This preparation evolved oxygen at more than 2000 μmol O_2 (mg Chl)$^{-1}$h^{-1}, when phenyl-*p*-benzoquinone was the electron acceptor, and it contained 60–80 Chl a per PSII.

Small but non-oxygen evolving PSII particles of current interest are, for example, the CP43 and CP47 complexes separated by Camm and Green (97, 135), after extraction by means of the detergent octylglucoside, and also the PSII preparation of Picaud et al in which digitonin-solubilised particles were resolved by means of electrophoresis in deoxycholate, and retained some PSII activity (136). Probably of greatest importance is the recent preparation, by means of Triton X-100, of the $D_1 D_2$ protein system by Satoh and Nanba (38) which carries out the reversible reduction of phaeophytin, and is not only the type material for the new hypothesis of PSII structure (see above) but also promises to bring to PSII the optical and experimental simplicity of the purple bacterial reaction centre.

Chapter 6

The Green Plant: Photosystem I

6.1 The Context of PSI

Photosystem I is a multi-subunit, chlorophyll–protein complex that exists in the thylakoid and carries out the photoreduction of ferredoxin by reduced plastocyanin. Plastocyanin, the electron donor (or cytochrome c-552 which replaces it in some algae), exists in solution in the lumen of the thylakoid, and the electron acceptor, ferredoxin, is in solution in the stroma. Photosystem I therefore drives an electron through the thylakoid membrane from the E to the P side. The action of the active centre, P700, was first reported by B. Kok, who observed the bleaching at 705 nm (137), later shown to be due to the photo-oxidation of a specific chlorophyll molecule, corresponding to the symbol C in Section 3.3.1.

6.1.1 Cyclic photophosphorylation

Unlike PSII, PSI can function on its own, by means of a cyclic electron transport pathway involving the bf complex. This provides a means of generating ATP and is known as cyclic photophosphorylation, because there is no change in any other reactant or product such as O_2 or $NADP^+$. The observation of the production of ATP, in 1954 by D.I. Arnon's group at Berkeley (138) brought in the novel concept that the chloroplast might be self-sufficient in photosynthesis. Production of ATP in this way was proved to be driven by PSI only, because it was not affected by the PSII inhibitor diuron, and also because the phosphorylation continued when the illumination was provided at wavelengths in excess of 700 nm (which do not allow PSII to work). Some processes that can be observed in intact algal cells, such as the uptake of glucose or potassium ions, require ATP, and can be shown to require light; the light is effective whether or not it is of more than 700 nm wavelength, so it can be concluded that, at least in these cells, cyclic photophosphorylation is available to increase the ATP supply when it is needed under any circumstance when NADPH is not limiting.

Cyclic photophosphorylation is possibly necessary in all photosynthetic cells, but it is the only possible function of PSI in cells where PSII is missing or inactive. Examples of such cells are the heterocysts of cyanobacteria, which use ATP for nitrogen fixation, and bundle-sheath cells of some C4 plants (see Chapter 11).

When PSI functions in tandem with PSII, the reduced ferredoxin is used to reduce (mainly) $NADP^+$, the coenzyme needed for the reduction of the metabolites of the carbon dioxide fixation pathway, although a number of other functions of reduced ferredoxin must not be overlooked. PSI is the eventual oxidant of the PQH_2 reduced by PSII, and in this case (known as non-cyclic electron transport) the products of the combined photosystems are O_2 and NADPH. In both the cyclic and non-cyclic processes, hydrogen ions are pumped into the thylakoid lumen during electron transport, and the consequent flow of protons through the coupling factor (F-ATPase) produces ATP; non-cyclic photophosphorylation was reported in 1958 by Arnon's group at Berkeley (139). The same group reported the use of the reduced dye $DCPIPH_2$ as an electron donor for the reduction of $NADP^+$ in 1961, and this readily performed reaction has served to characterise PSI activity in the same way that the Hill reaction characterises PSII.

The historical development of the two-light-reaction hypothesis was briefly indicated in Chapter 5.

Photophosphorylation of either kind depends on the translocation of protons during electron transport. In the non-cyclic pathway, for every electron passing from water to $NADP^+$, hydrogen ions are accounted for by (i) one left in the lumen by the oxidation of H_2O to O_2, (ii) one taken up from the stroma by the reduction of PQ at the Q_B-site (see Chapter 5), (iii) one injected into the lumen by the oxidation of PQH_2 at the *bf* complex, (iv) one (debatably) translocated through the *bf* complex from stroma to lumen by other means, such as by the Q-cycle (see Chapter 3) and (v) one consumed in the stroma by the metabolism of the reductive pentose cycle. This amounts to $3H^+/e^-$, or a P/O ratio ($P/2e^-$ ratio) of 2 if the ATPase is assumed to produce 1 ATP for $3 H^+$.

P/O ratios as high as this are seldom observed, and the usual figures of about 1.33 are consistent with a lack of the Q-cycle in the non-cyclic electron transport pathway, that is, with $2H^+/e^-$ instead of $3H^+/e^-$. This is an important issue, because the reductive pentose cycle that fixes carbon dioxide (Chapter 9) requires 3 moles of ATP and 2 of $NADP^+$ per mole of CO_2 fixed, that is, a P/O ratio of 1.5. If the Q-cycle does not operate, some contribution from cyclic photophosphorylation will be required to make up the deficit, not only for the reductive pentose cycle, but also for the formation of starch or sucrose, or other cell activities that require ATP. Jones and Whitmarsh (55) were able to show that 1.7–2.0 H^+ were injected into the thylakoid lumen, per electron transported from PQH_2 to PSI, at low flash frequencies, and they regarded this finding as evidence for a Q-cycle, but they also considered it possible that the mechanism might change to yield a lower H^+ ratio if a higher frequency of flashes, or steady light, were used.

Cyclic phosphorylation is sensitive to inhibition by the PQ analogue DBMIB (Fig. 6.1), and therefore involves PQ. It is also sensitive to potassium cyanide (KCN) (an inhibitor of plastocyanin), and therefore the electron pathway passes through the bf complex. The antibiotic antimycin A, which in bacteria and mitochondria inhibits the electron flow in cytochrome b, inhibits the cyclic pathway in chloroplasts, but does not inhibit the turnover of cytochrome b_6 in non-cyclic electron transport. Therefore in cyclic electron transport it is thought likely to be inhibiting the reduction of PQ by ferredoxin (see below).

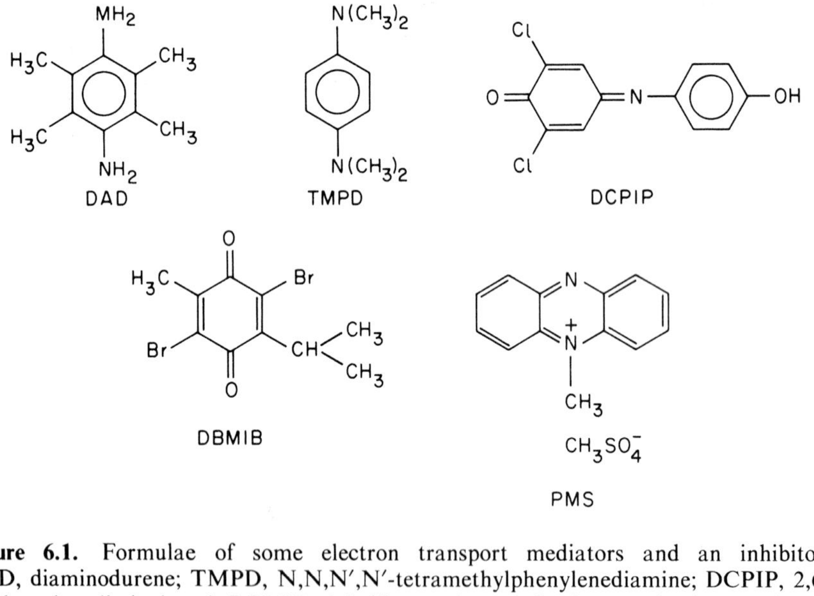

Figure 6.1. Formulae of some electron transport mediators and an inhibitor. DAD, diaminodurene; TMPD, N,N,N′,N′-tetramethylphenylenediamine; DCPIP, 2,6-dichlorophenolindophenol; DBMIB, 2,5-dibromo-3-methyl-6-isopropyl-p-benzoquinone; PMS, phenazine methosulphate

Cyclic phosphorylation can be mediated by artificial reagents in the absence of ferredoxin, such as diaminodurene (DAD) and phenazine methosulphate (PMS) (Fig. 6.1). These reagents react with and reduce plastocyanin and $P700^+$ respectively, and are themselves reduced by the PSI particle. The section of the natural photosynthetic electron pathway used in these cases was so short that it was a problem how it could be coupled to ATP formation; the essential point was that the reduced forms of these reagents ($DADH_2$, $PMSH_2$) were able to cross the thylakoid membrane and reduce the plastocyanin or $P700^+$ from the inside. The oxidised forms re-crossed the membrane to be reduced on the stromal side, and hence formed a proton-transporting system (see Fig. 6.2). The same process operates with the artificial electron donor system ascorbate–DCPIP; DCPIP (Fig. 6.1b) crosses the membrane and reacts with plastocyanin (140). Higher concentrations of these weakly acidic, lipophilic molecules, especially DCPIP, cause them to shuttle protons across the thylakoid membrane. This eliminates the PMF (Chapter 3) and hence they act as uncouplers of phosphorylation.

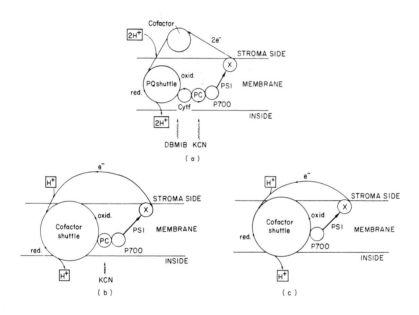

Figure 6.2. Mechanisms of energy conservation in cyclic electron transport. In case (a), the natural process where the cofactor is ferredoxin; electron transport is inhibited by cyanide or DBMIB and protons are translocated across the membrane as shown. In cases (b) and (c) artificial cofactors are used, for example DCPIP (or DAD), and PMS respectively. The cofactor itself forms a means of translocating protons. Reproduced from Trebst (140) with permission

Two einsteins of absorbed light are required per mole ATP formed, so taking the H^+/ATP requirement as 3 suggests that 1.5 H^+ are transported during the passage of 1 electron through the *bf* complex. The Q-cycle or some similar system would be needed.

It is uncertain how electrons from ferredoxin reach the *bf* complex. One possibility is that, just as in the non-cyclic scheme, ferredoxin reduces the flavoenzyme NADP reductase (FNR), which is tightly bound to the membrane, but nevertheless extrinsic, and FNR transfers 2 H^+ either to the PQ bound to the cytochrome *bf* complex or to mobile PQ in the membrane phase. The first possibility, of donation to PQ in the b_6f particle, is consistent with the FNR enzyme being virtually always found in preparations of the particle. On the other hand, analogy with mitochondrial complexes I and II (NADH:ubiquinone and succinate:ubiquinone reductases) would lead us to expect an independent particle acting as a ferredoxin:plastoquinone reductase (FQR). FNR is not a good candidate for FQR, since FQR (if it exists) is the site of the strong inhibition by antimycin A (to which the NADP reductase happens to be immune) (57). FQR may act on the PQ bound at the Q_c site of the *bf* particle, which is accessible to protons from the P aqueous phase, or on the pool of diffusible PQ in the membrane. To resolve this, new data may be needed on the behaviour of antimycin A.

There is a minority component of cyclic electron flow, which supports phosphorylation, which is not inhibited by antimycin A, and which works in broken chloroplasts that have lost their ferredoxin. It may involve a fraction of ferredoxin that is strongly bound (58), and work by direct contact between PSI and the *bf* complex.

Non-cyclic electron flow takes precedence over the cyclic process, presumably because in the presence of $NADP^+$, FNR preferentially reduces the coenzyme.

West and Wiskich (141) showed that broken chloroplasts showed 'photosynthetic control', in that the rate of the Hill reaction with ferricyanide was accelerated some two-fold when ADP was added (in the presence of inorganic phosphate and magnesium ions), and slowed when the ADP had all been converted to ATP. In the language applied to mitochondria, the chloroplasts were showing state 3/state 4 transitions, and this indicated tight coupling. However, there is evidence that the production of ATP can be diminished in intact chloroplasts without slowing electron transport, that is, coupling can be loosened. It was shown that isolated intact chloroplasts will carry out photoconversions of phosphoglycerate (PGA) to triose phosphate and of oxaloacetate (OAA) to malate. Both processes require one molecule of NADPH, and the first but not the second needs ATP. Production of NADPH ought to be slower by a factor of two or more when ATP is not being used, and therefore all the ADP is in the form of ATP. Put in another way, the reduction of PGA regenerates ADP and is a state 3 process, while the reduction of OAA does not, and should proceed at the slower rate characteristic of state 4. In fact both reactions were observed to proceed at approximately the same rate. The quantum requirements were four in each case (142), and flexible coupling was the suggested explanation.

6.2 Isolation of PSI

Photosystem I was first isolated from a digitonin digest (in high salt) of chloroplast thylakoids followed by differential centrifugation, by Boardman and Anderson (129). This preparation sedimented at 100,000 g, and consisted of vesicles from stroma thylakoids, with their natural preponderance of PSI.

A smaller particle results from digesting thylakoids with higher concentrations of digitonin, in low salt. These are readily separated by means of density-gradient centrifugation into bands containing LHCII, PSII and PSI in increasing order of density (143) (Fig. 6.3). The membrane lipids can be removed from the vesicles by means of detergents such as LDAO, octylglucoside or dodecylmaltoside, yielding chlorophyll–protein particles that can be resolved in polyacrylamide gel electrophoresis. They are active in the DCPIPH$_2$–ferredoxin reaction.

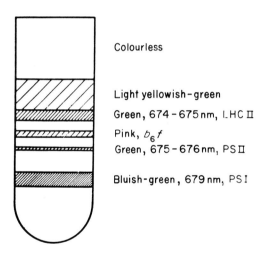

Colourless

Light yellowish-green

Green, 674-675 nm, LHC II

Pink, $b_6 f$

Green, 675-676 nm, PSII

Bluish-green, 679 nm, PSI

Figure 6.3. Separation of PSI and PSII particles by means of digitonin. At low salt concentrations PSII is not sedimented in a density gradient. Bands corresponding to LHCII, the $b_6 f$ complex, PSII and PSI can be seen. Reprinted by permission of Kluwer Academic Publishers. Copyright © 1972 by Dr W. Junk Publishers. From Wessels, J.S.C. & Voorn, G. (1972). In G. Forti et al, eds, *Proc. 2nd Int. Cong. Photosynth. Res.*, pp. 833–845

Application of the detergent Triton X-100 in high concentration yields similar particles, TSF-1 of Vernon and Shaw (144). In both preparations, the ratio of chlorophyll to P700 was about 100:1. A particle with a 30:1 ratio was obtained with the same procedure from freeze-dried carotene-extracted chloroplasts, called HP-700 ('high P700 particles'). The Triton particles do not lend themselves to electrophoresis because the detergent molecules do not carry any charge, but it would appear from analyses of the peptides present (using SDS which is negatively charged) that the LHCI antenna and some chlorophyll from the core complex has been removed by the Triton procedure.

Smaller chlorophyll–protein complexes were obtained by extracting thylakoid material with the detergent SDS, followed by electrophoretic separation on polyacrylamide gel. If the detergent is added at room temperature (SDS is relatively insoluble in the cold) three chlorophyll-containing bands are obtained (Fig. 6.4(i)), labelled by Thornber et al (145), in increasing order of mobility, complex 1, complex 2 and free pigment that was later shown to have arisen from the destruction of PSII. Later the abbreviations CP1 and CP2 were employed. This application of SDS is known as 'denaturing conditions', but it must not be confused with the denaturing conditions used, for example, in the estimation of peptide masses on polyacrylamide gels when the sample is boiled with SDS solution. CP1 was shown to be absent from a PSI-deficient mutant of *Scenedesmus*, suggesting that it could be identified with (part of) PSI; the cyanobacterium *Phormidium luridum* yielded a CP1 with detectable P700 activity, confirming the identification. CP1 from higher plants does not readily show P700 activity, because the iron–sulphur acceptors are destroyed by SDS, and the charge-separated couple produced at the reaction centre recombines very quickly.

At the outset of the 'detergent era' of thylakoid analysis, it was feared that chlorophyll could be extracted from one site, dissolved as a chlorophyll–detergent micelle, and become adventitiously attached to other sites or other proteins, producing false chlorophyll–protein complexes. Chlorophyll is certainly extracted, totally in the case of SDS and PSII, but reassociation has never been observed. The technique of CD readily reveals changes in the relative positions of the chlorophyll molecules accompanying (partial) denaturation of chlorophyll–protein complexes, and there is now much less fear that artefactual complexes are produced. Improvements in the application of detergents (for example using the lithium version of SDS, LiDS, in the cold) increase the number of bands seen in gel electrophoresis, and these can to some extent be interpreted as combinations and oligomers of the PSI core complex, LHCI, the PSI antenna, PSII and LHCII. The free-pigment zone is greatly suppressed and there is no reason to believe that free pigment (not bound to protein) exists in the thylakoid (Fig. 6.4 (ii)–(viii)).

CP1 as first obtained on polyacrylamide gel electrophoresis is a dimer or hetero-oligomer of two related but non-identical peptides of some 66 kDa each, which with one P700 (but see below) probably represents the unit-form of the reaction centre. Thornber (14) points out that estimates of molecular mass under these conditions are very unreliable. There are two genes coding for the two peptides, *Psa*A and *Psa*B (146), in the chloroplast genome, although that does not rule out their products being interchangeable: the multiple genes for LHCII seem to have interchangeable products.

Owing to the proliferation of chlorophyll–peptide complexes with CP desig-

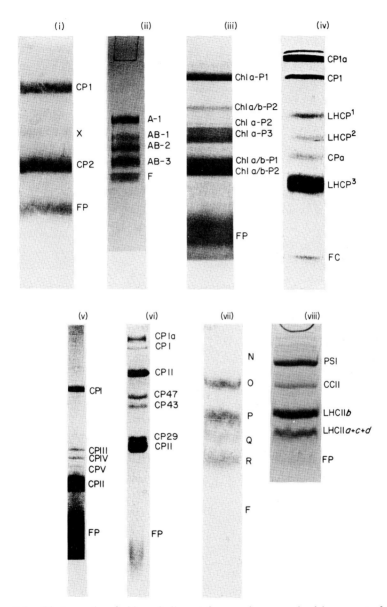

Figure 6.4. Photographs of chlorophyll–protein complexes resolved by means of poly-acrylamide gel electrophoresis, in eight different procedures. The dark bands are the chlorophyll-containing components: the gels are not stained and non-pigmented proteins are not visible. (i)–(iv) SDS in the gel (250–253); (v) LiDS in gel, at 4°C (253); (vi) octyl glucoside/SDS (254); (vii) Deriphat-160, a zwitterionic detergent, in the gel (255); (viii) extract made with glycosidic surfactant and run in Deriphat. (i)–(vii) reproduced with permission; (viii) by courtesy of G.F. Peter and J.P. Thornber

nations, it was suggested that functional complexes should be labelled CCI, CCII for core complexes, and LHCI, LHCII for antennae. On this basis CP1 is the major part of, and the only chlorophyll–protein complex in, CCI. The number of pigment molecules in CCI is not precisely known, but is commonly put at 90 Chl a and at least 2 β-carotene per P700, the previous estimate of 40 Chl a (assuming one P700) being now regarded as low (14). The number allocated to each 66 kDa peptide, of course, depends on the number of such peptides assumed to be present in CCI, and whether they have equal chlorophyll complements. Nevertheless it is striking that the mass of the chlorophyll present is close to one-third of the mass of the protein binding it, a very high proportion compared to virtually all other pigment–protein complexes known.

The 66 kDa peptides are membrane-spanning intrinsic components of CCI, and there are associated with them several extrinsic, small peptides. The 66 kDa

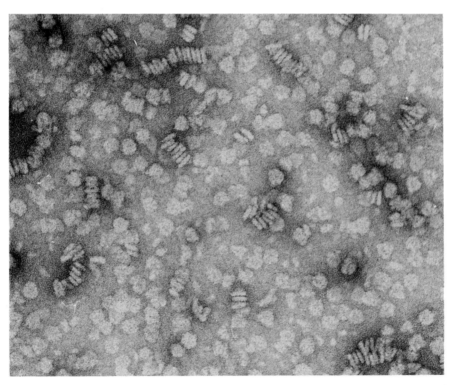

Figure 6.5. Electron micrograph of a trimeric PSI particle. The PSI particle prepared from the cyanobacterium *Synechococcus* sp. has the form of a disc that can be resolved into three sector-shaped subunits. The diameter is 19 nm and they are 6 nm thick. The complex has a mass of 600 kDa and 120 Chl a molecules per active P700 centre. Prepared in *n*-octyl–D-glucopyranoside and negatively stained with 1% uranyl acetate. Reproduced by permission of Dr E.J. Boekema and Elsevier Science Publishers. From Boekema, E.J. et al (1987) *FEBS Lett.* **217**: 285

(chloroplast encoded) peptides above have been termed 'subunit I'; CCI also contains nuclear encoded, non-chlorophyll-containing peptides. Of these, subunit II (25 kDa in chloroplasts, 16 kDa in cyanobacteria) is suggested to assist in the assembly of CCI, subunit III (20 kDa) is believed to assist the binding of plastocyanin, and up to four smaller peptides exist, down to 8 kDa, of which at least some carry the iron–sulphur centres that form secondary electron acceptors. The association between polypeptides and gene-products, coded in the chloroplast or nuclear genomes, is shown in Table 13.1.

Boekema et al (147) isolated a PSI complex from *Synechococcus* sp., using the material left after the extraction of PSII by means of the detergent SB12 (see Section 5.8); application of β-dodecylmaltoside and density-gradient centrifugation yielded a particle that appeared trimeric in electron micrographs (Fig. 6.5), and had a mass of approximately 600 kDa by gel-filtration. The monomers (230 kDa) could contain the two peptides of about 66 kDa, some low-mass proteins and pigments each, but there was no evidence that the monomers were identical, nor how many P700 centres there were in the complex. The Chl a : P700 ratio was 120:1, and a substantial part of the LCHI antenna complex could be assumed to be present.

As indicated in Chapter 2, the chlorophylls are differentiated by their local protein environments into groups with specific wavelength maxima of absorption and fluorescence emission. These groups do not appear to be correlated with the 66 kDa proteins, so that it has not proved possible to extract a reaction-centre complex as simple as that of the purple bacteria. Nevertheless it appears that the outer chlorophyll of the complex can be detached, for example by the use of the detergent LDAO, and active P700 preparations can be obtained with as few as 10–12 Chl a molecules (148) or 20 Chl a (149).

6.3 Electron Transport in the PSI Complex

6.3.1 The reaction centre

Following the standard pattern for reaction centres (see Chapter 3), electron transport is driven by the excited state of a special Chl a (or derivative) monomer or dimer; it donates an electron to a primary acceptor, leaving the radical cation which can be detected by a bleaching at 700 nm. This provides a simple spectrophotometric assay for PSI preparations. The active pigment is known as P700. The radical cation P700$^+$ can also be generated chemically, by oxidising agents such as potassium ferricyanide $(K_3Fe(CN)_6)$ redox buffers, and has been shown to be a pH invariant one-electron couple with $E_m = 490$ mV, a value which has been revised upwards (150) from Kok's earlier determinations (151). The excited state P* is efficiently quenched by photochemistry, and has no identified fluorescence; P700$^+$ is a quencher of excitation in PSI, resulting in the low level of fluorescence from PSI at 25°C.

$P700^+$ has a characteristic ESR spectrum (g=2.0025, width 7.2G). The width is narrower than for $(Chl\ a)^+$ generated in vitro (9.3G), and this was attributed to a dimer structure for P700, so that the electron was delocalised over two molecules. A dimer structure is of course established for the reaction centre of purple bacteria, and hence by analogy at least for PSII, but the analogy can hardly be pressed further. A pair of Chl a molecules having an exciton interaction was also suggested by Philipson et al (152) from CD data. Triplet-minus-singlet spectra have been recorded by Hoff (104) which match the in vitro dimer $(Chl\ a)_2$ better than the monomer in the cases of both PSI and PSII. At low temperature the $(P^+ - P)$ difference-absorption spectrum shows a feature (+ve at 695 nm, −ve at 685 nm) that could be interpreted as an electrochromic shift of a monomer Chl a close to P^+. Recent ESR and ENDOR measurements have produced a closer match with monomeric Chl a, but have also suggested that ring V (isocyclic) is modified in P700. If ring V were in the enol form, rather than the keto form as normal, it could account for the E_m of P700 being lower than that of Chl a (0.5 V rather than 0.86 V). It is not unreasonable to adopt the same position as for PSII, that P700 in the ground state involves two (or more) excitonically interacting Chl a molecules, but that the radical cation only involves one molecule.

The above discussion does not preclude the involvement of Chl a modified in additional ways. Dörnemann and Senger (153) have isolated a derivative (13^2-hydroxy-20-chloro-chlorophyll a) (Fig. 6.6a) from a great variety of PSI containing sources, and they point out that it is always present in a 1 : 1 proportion to P700, and may therefore be the actual molecule underlying the purely spectral phenomenon of P700. They have labelled this form Chl RCI, but it is not certain whether its identity with P700 is genuine. Suggestions that Chl RCI is an artefact are made by Watanabe et al (154) who have detected the 10-epimer of Chl a (Chl a', Fig. 6.6b), also in a proportion of 1:1 with P700. It must be noted that in all three hypotheses, (i) the ring V enol, (ii) the hydroxylated form at 13^2 and (iii) the 10-epimer (the same carbon on the old system of numbering), have much in common, and (ii) and (iii) could relatively easily arise from (i). Watanabe claims that the Cl-20 atom (on methene bridge δ) could have originated from contact of normal Chl a with, say, ordinary filter paper which has an appreciable chloride content, and it would not be difficult to envisage its formation from the enol form of Chl a.

6.3.2 Intrinsic acceptors

ESR at low temperature showed the existence of an electron acceptor which had the properties of an iron–sulphur centre, and which was labelled 'bound ferredoxin'. At the same time optical absorption spectra were obtained by flash spectroscopy, and were labelled P430. The redox potentials of these forms were low, in the region of −0.5 V, but correlation was difficult. The ESR mea-

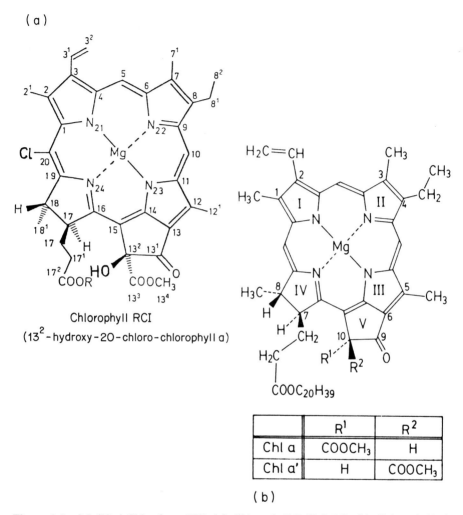

Figure 6.6. Modified Chl a from PSI. (a) Chlorophyll RCI (153). (b) Chlorophyll a′ (154). (Note that (a) illustrates the modern numbering system. In (b) the traditional numbering system has been used)

surements were refined (see the review by Evans (155)), and they showed the presence of two components, now labelled F_A and F_B, with potentials of -540 and -590 mV respectively (in spinach chloroplasts). They are identified as Fe_4S_4 clusters. At low temperature, illumination only results in one turnover of P700, and one electron enters the F_A–F_B system. Contrary to earlier reports, the electron appears to be shared between the two, which together constitute P430 and the ultimate electron acceptor in PSI.

Sequence studies on the 8 kDa peptide of CCI (coded by chloroplast gene *Frx*A) show the characteristic distribution of cysteine residues characteristic of bacterial (Fe_4S_4) ferredoxins (156), and the same peptide has been shown to be necessary for ferredoxin reduction (157). This peptide is a likely candidate for at least one of the P430 iron–sulphur centres.

If the centres F_A and F_B in a PSI preparation are chemically pre-reduced, and then illuminated at low temperature, another Fe_4S_4 centre becomes reduced, at a potential of -705 mV, and is labelled F_X. Studies by flash-ESR and absorption discounted F_X as an intermediate at low temperature. However, experiments by Sauer et al (158) had shown the existence of a redox intermediate labelled A_2, which was reduced at room temperature by one flash when P430 was reduced. The suggestion that F_X and A_2 were the same was confirmed by Sétif et al (159), so that this third Fe_4S_4 centre is an efficient intermediate at normal temperatures. There are 12 Fe and 12 S to be accounted for in the PSI reaction centre, and the above assignment to F_A, F_B and F_X of Fe_4S_4 would each appear to be exhaustive.

It has been stated above that iron–sulphur centres are destroyed in higher-plant PSI by the detergent sodium dodecyl sulphate. This allows the reduction of an earlier acceptor, labelled A_1, to be observed. In the same kind of experiment ESR studies showed the formation at low temperature of the triplet form of P700, and the relationship of the triplet with the putative electron acceptors led to the postulation of an even earlier acceptor A_0 (as well as A_1); the triplet is presumed to be formed by the back-reaction of $[P700^+\text{-}A_1^-]$. Picosecond spectroscopy by Shuvalov (160) led him to suggest that A_0 was a Chl a molecule (absorbing at 693 nm). Low-temperature flash difference-spectroscopy of A_1 showed optical and ESR characteristics of a quinone–semiquinone transition. Given the occurrence in PSI of phylloquinone (vitamin K_1) (Fig. 6.7), this substance is provisionally identified as A_1. On the other hand the phylloquinone, of which there are two molecules in PSI, could be almost totally destroyed by UV irradiation without altering the shape and size of the ESR spectrum of A_1, or the photo-oxidation of P700 (161), casting some doubt on the K_1–A_1 identification; the question is still open.

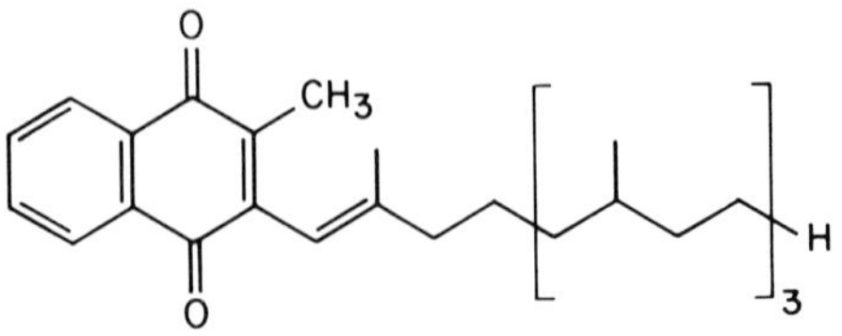

Figure 6.7. Phylloquinone (vitamin K_1)

6.4 External PSI Electron Transport

6.4.1 The donor side

P700$^+$ is re-reduced by plastocyanin, a protein of M_r 10,500, which contains one atom of copper. It was discovered by S. Katoh (162), and has been intensively studied by X-ray crystallography. Figure 6.8 shows the structure determined by Guss and Freeman at 0.16 nm resolution (163). It is a basket or sandwich structure of β-strands, containing the copper ion (Cu^{2+}) bound to His 37, Cys 84, His 87 and S-met 92, in a very distorted tetrahedral structure that changes with pH. The redox potential is relatively high: $E_0' = 370$ mV (Cu$^+$/Cu^{2+}). All the hydrophobic amino acid residues are internal, and

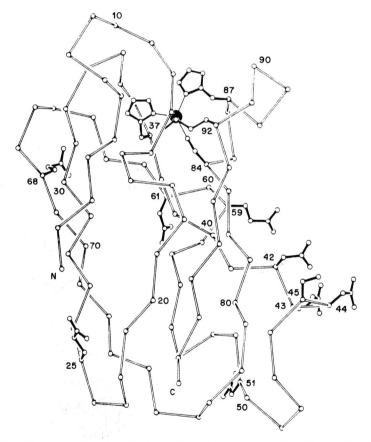

Figure 6.8. Structure of poplar plastocyanin at 0.16 nm resolution. The peptide backbone (α-carbons only) is shown, with important residues such as the ligands to Cu (towards the top) added. Reproduced by permission of Dr J.M. Guss, Professor H.C. Freeman and Academic Press. From Guss, J.M. & Freeman, H.C. (1983) *J. Mol. Biol.* **169**: 550

plastocyanin is very soluble. Amino acid sequences are known for many species, and 64 out of the 99 amino acids are conserved (see the review by Haehnel (164)).

Histidine-87, a ligand to the copper, is exposed to the solvent, and is a possible site for the conduction of the electron during redox reactions. Alternatively there is a chain of four aromatic residues extending from the copper to the opposite surface. Thirdly, there is an acidic patch on the surface, made up of six (Asp, Glu) residues at positions 42–45, 59 and 61, which could serve as a recognition site for either P700 or cytochrome f, with Tyr 83 in a position to act as a possible electron channel. Although there is discussion on the point, there is no good indication of the formation of a complex between PSI and plastocyanin, and hence it has been difficult to ascertain whether molecular recognition sites exist. (The possible involvement of subunit III of PSI in binding plastocyanin, mentioned above, may provide a lead.) The same is true of the reaction between plastocyanin and cytochrome f. Nevertheless both of the reactions of plastocyanin are very fast. The speed is surprising since the narrowness of the thylakoid lumen must restrict the diffusion (with respect to diffusion measured in vitro). The molar proportions of plastocyanin to P700 and cytochrome f are $2:1:1$.

At low pH values, the His 87 imidazole group becomes protonated (pK 5.4) and moves away, leaving trigonally bound copper which is relatively stabilised in the Cu^+ form. This might be expected to slow the reaction between reduced plastocyanin and $P700^+$ and provide a means of regulating electron transport in response to the transmembrane pH gradient (that is, in state 4) as a supplement to the currently accepted pH-regulation by the PQ–Rieska protein reaction.

Plastocyanin is coded in the nucleus, and the gene has been sequenced in several species.

In some algae (e.g. *Chlamydomonas*, *Scenedesmus*) copper deficiency results in the replacement of plastocyanin by a soluble cytochrome c-552.

Plastocyanin is lost from the lumina of thylakoids during sonication and must be re-added if PSI activity is to be observed; there is no counterpart to the claimed ability of some bacteria to dispense with their soluble cytochrome c (81), (see p. 96).

Diffusion of plastocyanin is as important and difficult a question as the diffusion of PQ. According to a determination of the lateral distribution of PSII, PSI and the bf complex in thylakoids, by Morrissey et al (165), PSII was found in the grana thylakoids and PSI in the stroma thylakoids, as expected, and the bf complex was associated with both in a way that suggested that it belonged to the 'fret' that connects the granum disc with the rest of the (stroma) thylakoid. Seen from the edge, the distance diffused by PQ from PSII to the bf complex would be similar to that diffused by plastocyanin from the bf complex to PSI. The lumen of the thylakoid appears narrow enough in micrographs, even assuming that all the membrane material stains well with electron-dense stain, and it must offer considerable resistance, a topic discussed by Whitmarsh (166).

Cytochrome f is part of the b_6f complex, replacing cytochrome c_1 of bacteria and mitochondria, with which it has little common primary structure. Analysis of the gene (chloroplast: petA) suggests that the 31 kDa protein is anchored in the membrane by a single α-helical segment, leaving the bulk projecting into the lumen of the thylakoid. There is no structural model for the hydrophilic region, which includes the haem-binding region, but a general model for the bf complex is shown in Fig. 6.9 (167) which shows the cytochrome f haem adjacent to the iron–sulphur Rieske cluster in the luminal space. The pH-independent mid-point potential is 340 mV, and the cytochrome is easily monitored by its α-band for the reduced state at 553 nm (algae), 554 nm (chloroplasts) and 556 nm (cyanobacteria). Both the haem of cytochrome f (in the model) and the copper of plastocyanin are inside their protein domains, and the question of the electron transfer path is acute. There is a concept of 'electron tunnelling' according to which an electron can pass from one region to another of similar or more positive potential even when there is a barrier of high negative potential between, making use of the indefinitely extended probability field of the electron. The penalty is speed, long tunnels being intrinsically slow; it has been mentioned above that plastocyanin reactions are very fast. The chemistry of electron transfer between inorganic centres in these proteins is therefore far from being explained.

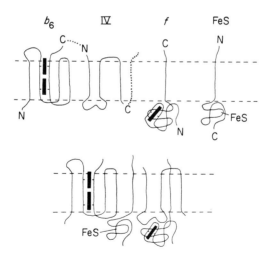

Figure 6.9. Proposed topographical arrangement of the b_6f polypeptides in the membrane. In the upper part of the diagram, the membrane-spanning and soluble parts of the various polypeptides are shown separately. Haem is represented by heavy bars. The location of the N-terminus of cytochrome b_6 is arbitrary. In the lower part of the Figure the polypeptides are arranged as they may be in situ. Reproduced from Hauska (167) by permission of Springer-Verlag

6.4.2 The acceptor side

So far as is known, the only acceptor for electrons from the terminal acceptors of PSI (P430) is ferredoxin. Chloroplast ferredoxins differ from their bacterial counterparts in their larger molecular mass (10,500–11,000 kDa), redox potential (-420 mV at pH 7) and electron capacity (one electron as opposed to two). They are deep red-brown in the oxidised state, and are substantially decolorised on reduction. There are some sequence homologies between the plant and bacterial series, suggesting that a series of doublings has taken place in the evolution of plant type from presumed ancestral bacterial type ferredoxins. All the plant types are very similar, some 46 out of 98 residues being strongly conserved.

In conditions of iron deficiency, some green algae, purple bacteria and cyanobacteria have been observed to form a protein known as flavodoxin (old name: phytoflavin), which substitutes for ferredoxin in all its activities. Flavodoxin has a mass of 20 kDa and one molecule of flavin mononucleotide (FMN).

There is no evidence for the involvement of any other PSI peptide in the transfer of electrons from P430 to ferredoxin, although it is not known which of the minor peptides in CCI actually carries either or both of the iron–sulphur clusters that constitute P430.

Reduction of $NADP^+$

The enzyme ferredoxin:NADP reductase (FNR: EC 1.6.99.4) is normally firmly bound to the stromal side of the non-appressed thylakoids, and requires prolonged washing to remove it. It is not known where it binds, but it is commonly found as a constituent of the *bf* complex. Removal from the membrane diminishes its affinity for ferredoxin. The protein is a single peptide of mass 34 kDa with one molecule of flavin adenine dinucleotide (FAD). The reduced flavin transfers H^- (hydride) to $NADP^+$. It can be assayed spectrophotometrically by a diaphorase reaction in which the blue indophenol dye DCPIP is reduced by NADPH.

Besides its role in the reduction of FNR, ferredoxin is also required for the reduction of nitrite (to form ammonium ions), sulphate (forming sulphide), thioredoxin (p. 164) and oxoacid synthases (p. 167). The expected reaction with oxygen to generate superoxide (O_2^-) is only observed in vitro; presumably the efficient superoxide dismutase (EC 1.15.1.1) (or ascorbate, which reduces superoxide to peroxide) and catalase are effective scavengers. However, certain compounds such as naturally ocurring phenols and the artificial viologen reagents are reduced by P430 in experimental conditions and then react with oxygen, producing superoxide or hydrogen peroxide. This results in a net light-dependent oxygen uptake (the Mehler reaction), which is eliminated by the addition of catalase and superoxide dismutase, leaving an apparently cyclic system in which oxygen has a catalytic role, known as pseudocyclic electron transport (or pseudocyclic photophosphorylation).

6.5 Summary: PSI, PSII and the Z-scheme

By way of a summary of Chapters 5 and 6 together, Fig. 6.10 is the traditional Z-scheme, which shows the pathway of an electron from water through all the known or possible redox carriers to the acceptor $NADP^+$. The scheme uses the vertical scale to indicate the redox potential of the carriers at each stage; it is not known in most cases what the redox balance is at any given point in steady-state conditions, and the potentials shown are standard ones. It would be rash to identify any part of this scheme as absolutely certain, but three regions particularly are the subject of continuing active research and debate: the donors to PSII (S, Z and D), the acceptors from PSI (A_0, A_1 and the iron–sulphur centres), and the features of the b_6f particle (such as the Q-cycle that allow the additional proton pumping).

140

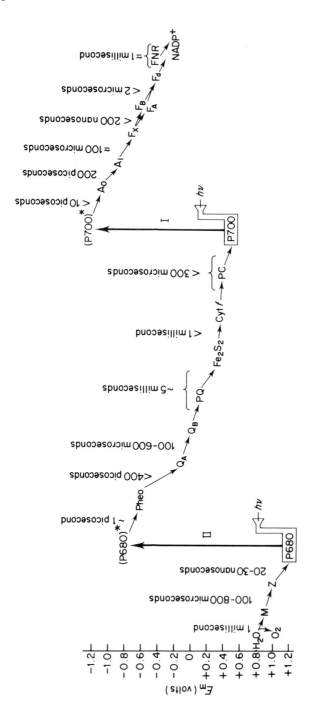

Figure 6.10. The Z-scheme: pathway of non-cyclic electron transport in chloroplasts. The ordinate represents the standard redox potential of the intermediates represented along the abscissa, so that the electron flow is from left to right. The characteristic reaction times are indicated for each step. Symbols and abbreviations are as used in the text, except that M is an all-purpose complex including the components of the oxygen-evolving system. The Z-scheme originates from Hill & Bendall (88); this modern version is reprinted by permission of Kluwer Academic Publishers. Copyright © 1986 by Martinus Nijhoff Publishers. From Govindjee & Eaton-Rye, J.J. (1986) *Photosynthesis Res.* **10**: 366.

Chapter 7

Green Bacteria; Summary of Photosynthetic Electron Transport

7.1 The Status of the Green Bacteria

The colour of the green bacteria is due to their complement of BChl c (bacterioviridin, Chlorobium Chl 660) or BChl d or e, which is contained in the characteristic antenna structures termed chlorosomes, described in Chapter 2. Much less has been done with this group, partly because they present greater problems of culture (most species require complex media or mixed cultures) and are very sensitive to the presence of oxygen.

The 'original' green bacteria, the Chlorobiaceae, known as the green sulphur bacteria in contrast to the purple sulphur bacteria represented by, for example, *Chromatium*, included the laboratory species *Chlorobium thiosulphatophilum*, *Cb. limicola* and *Chloropseudomonas ethylica*. After some confusion in the cultures, *Prosthecochloris aestuarii* was identified with, and replaces, the former *Cp. ethylica*, and the name *Cb. limicola* f.sp. *thiosulphatophilum* replaces the other two. There is therefore no distinction between *Chlorobium* and *Chloropseudomonas*. It is advisable in consequence to treat with reserve species named as *Cp. ethylica* or *Cb. limicola* in work done in the period before 1974–6 when the taxonomy was sorted out. *P. aestuarii* is now the most studied of the group.

They are all obligate anaerobes, and require sulphur compounds to provide the reducing equivalents for carbon dioxide fixation (by whatever means; see Section 9.6.2). *Cp. ethylica* was supposed to be able to use an organic reductant (preferably C2 such as ethanol) which did not necessarily have to enter the metabolic pool (a lifestyle termed myxotrophy), but it was established that the C2 compound was in fact required by a symbiotic species of non-photosynthetic bacterium growing in the old culture.

The *Chloroflexus* group, discovered recently by Pierson and Castenholz (168), are facultative aerobes. They are filamentous in form, and they have a gliding behaviour in which they resemble cyanobacteria, or non-photosynthetic gliding bacteria. They occur in algal mats in hot springs, under conditions in which they are aerobic and largely de-pigmented, so that they were hard to recognise as photosynthetic bacteria until they had been cultured axenically and anaerobically. They can be referred to as the green non-sulphur bacteria, in contrast to the green sulphur bacteria described above. Although the *Chloroflexus* group resemble the *Prosthecochloris* group in the possession of BChl c and the chlorosome structure, they resemble the purple bacteria in the structure of their reaction centre. A study of the ribosomal RNA sequences (169) revealed no basis for considering the two groups of green bacteria to be particularly related.

The traditional method for discovering photosynthetic bacteria is to take samples of pond mud and keep them in deep glass vessels in the light, with or without provision of sulphide. The mud quickly becomes anaerobic and coloured colonies of bacteria develop against the glass. By this means a new species was discovered, named *Heliobacterium chlorum*, whose relationships are not yet clear. It contains a novel pigment, BChl g (see Chapter 2), and is included in this section without prejudice to its final assignment.

7.2 Structures

The chlorosomes (Figs 2.8, 2.9) of the two groups of green bacteria are similar in size and shape, and are both composed of rod elements which differ only in minor details. The rods in each case are connected to a base-plate, and there is here more variation; the base-plate of Chlorobiaceae is a water-soluble complex that has been crystallised by Matthews and co-workers (9) and is unrelated to the base-plate proteins of the Chloroflexaceae. The base-plate is attached to the membrane by (in the Chlorobiaceae) large (12 nm) particles or (in the Chloroflexaceae) a crystalline array of 5 nm particles. Underlying the attachment sites are the reaction centres.

7.3 Reaction Centres

Like the purple bacteria, each of the two groups of green bacteria has only one kind of reaction centre.

In the green non-sulphur bacteria the pattern is little different from that in the purple bacteria, except that the H subunit of the reaction centre is missing and there are 3 BChl a and 3 BPh a molecules attached to the L and M subunits instead of 4 and 2 respectively in the purple bacteria. Studies of amino acid sequences have led to the proposition that *Chloroflexus* represents the most evolutionarily primitive form of photosynthetic organism. The photoactive BChl a is found as P870, the same as in *Rps sphaeroides*.

The photochemically active pigment in the green sulphur bacteria is BChl a as a dimer in P840 (in *Prosthecochloris* and *Cb. limicola* f. *thiosulfatophilum*). The sequence of electron acceptors is not at all well established. Studies by the groups of J.M. Olson (170) and of J. Amesz (171) are in a measure of agreement that the primary acceptor is a BPh (a or c) molecule, and that iron–sulphur centres are secondary acceptors. Amesz found an intermediate acceptor, BChl a (814 nm) (171).

The reaction-centre complex obtained by Hurt and Hauska (35) contained the 65 kDa bacteriochlorophyll polypeptide as well as cytochrome and a possible Rieske-type iron–sulphur protein, which might imply that this arguably primitive system combined the characters of a reaction centre with the bc complex. The existence of photoreducible iron–sulphur centres of the kind described by Amesz could not be confirmed (172), and in the same report ESR lines were described which were very similar to those of the acceptors A_0 and A_1 in chloroplast PSI.

The electron donor to $P840^+$ is cytochrome c (c-553). The relatively slow reaction time of 90 μs indicates a loose attachment if any. The tentative identification of a bacteriophaeophytin molecule is analogous to the purple bacterial reaction centre, but the (alleged) iron–sulphur centres are more strongly analogous to green-plant PSI, as is the size of the principal peptide, 65 kDa, that carries the BChl a. The cytochrome c donor to $P840^+$ could be related to either group of reaction centre, remembering that a c-cytochrome can substitute for plastocyanin in algae. The BChl a acceptor resembles A_0 in the chloroplast PSI. Figure 7.1 compares the electron transport processes of the purple bacteria with those of the two groups of green bacteria.

Although the presence of iron–sulphur proteins has not been confirmed in either *P. aestuarii* or *Cb. limicola* there is evidence for iron–sulphur centres in *Heliobacterium chlorum*. Data on this new species is scanty, but cytochrome c-552 has been identified as the donor to the reaction centre (P798, a BChl g dimer), and an early electron acceptor resembles BChl c (174).

Reduction of ferredoxin in green sulphur bacteria is required for fixation of carbon dioxide by the reductive citrate pathway. NADH is also required, and if, as has been claimed, RuBisCO (see Chapter 9) is present in strains of *Cb. thiosulfatophilum*, NADH will be needed for the reductive pentose cycle also. A ferredoxin:NAD reductase has been reported on the inner (P) face of the cell membrane.

There is also a lack of detail concerning the electron transport from the inorganic reductant to the cytochrome donor to P840. Menaquinone is present, and cytochromes b and c have been identified in active particles (35), consistent

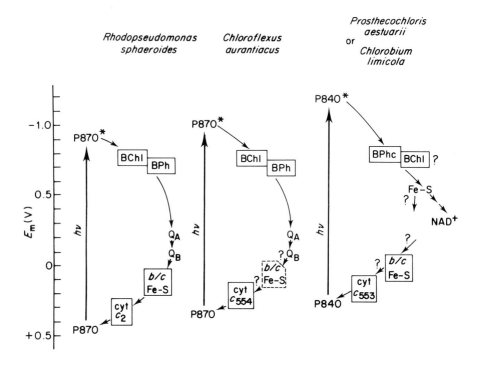

Figure 7.1.   Electron transport pathways in purple and green bacteria. The diagrams are based on the same principle as Fig. 6.10. The green bacterium *Chloroflexus* is clearly very similar to the athiorhodacean *Rps sphaeroides*, but differs significantly from the Chlorobiaceae. There is no evidence for the involvement of menaquinone, but it is an interesting possibility, in the green sulphur bacteria. Reproduced from Blankenship & Fuller (173) by permission of Springer-Verlag

with the earlier report (170) of an ESR signal attributed to a Rieske centre. The status of the cytochrome *bc* complex is therefore reasonable (but not established), and the mechanisms described for the purple bacteria (Section 4.6) need not be very different here.

7.4 Comparison of Electron Transport Processes

Figures 7.2 to 7.8 show an arrangement of somewhat generalised redox carriers or particles that can provide a rationale for the various electron transport processes carried out by most organisms. In this form we can construct first the electron transport system of the mitochondrion (Fig. 7.2), noting that the electron donor system (AH_2–FP) represents complex II, containing the flavoprotein dehydrogenase, specific for NADH and succinate. The simple cyclic

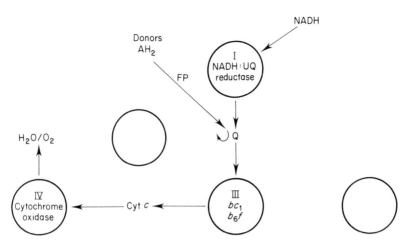

Figure 7.2. A sketch of mitochondrial oxidative electron transport, in a form for comparison with the photosynthetic types. In this case the 'donors AH_2', for example succinate, are connected to the ubiquinone pool by specific flavoprotein-containing complexes such as complex II

light-driven electron transport system of purple bacteria is represented by Fig. 7.3, to which we add the reverse electron transport (Fig. 7.4) sequence that is required to generate NAD^+.

The hydrogen donor in the non-sulphur bacteria would be like succinate, reducing the quinone via a flavoprotein complex without supporting hydrogen ion translocation. Although hydrogen sulphide has a low enough potential to reduce quinone directly (via an enzyme) in solution, no enzyme is known that carries out the reaction.

The photosystems of the green plant in the non-cyclic mode are represented by Fig. 7.5, and in the cyclic mode by Fig. 7.6.

The ambiguous path from ferredoxin to the *bc* complex in green-plant cyclic electron transport is yet to be resolved.

Some algae have cytochrome c_2 in place of plastocyanin, which allows a scheme similar to Fig. 7.6, and 7.7 to be proposed for the Chlorobiaceae, in cyclic and non-cyclic modes respectively. In these bacteria, however, the details need confirmation. The principal reason for supposing that a cycle like PSI operates in this case is that otherwise it is not obvious how the ATP is supplied. Less uncertain is the role of sulphur compounds, which provide electrons to the donors to the photosystem. Since the supply of environmental reduced sulphur and the cytochrome *c* are both on the same (E) side of the cell membrane (but the *c*-cytochrome is not unambiguously shown to be the soluble one), the entry of electrons from sulphur to the electron transport chain could be at the level of the cytochrome. This would not explain how the PMF for the ATP supply

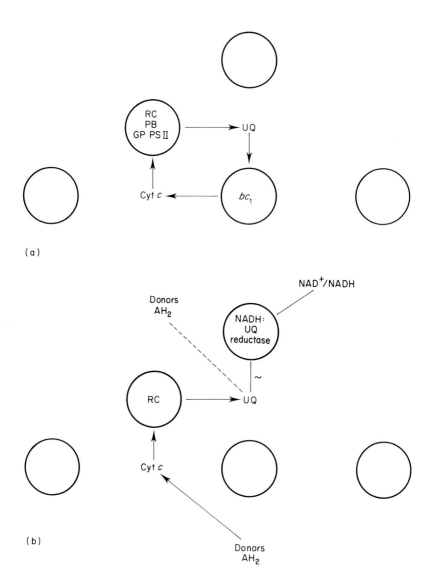

Figure 7.3. Photosynthetic electron transport in purple bacteria. (a) Cyclic electron transport involving the reaction centre and *bc* complexes. RC, photosynthetic reaction centre. PB, purple bacteria. GP PSII, green-plant: photosystem II. (b) Reverse electron transport, in which energy (in the form of PMF) is used to drive the reduction of NAD^+ by means of electron donors coupled via ubiquinone or cytochromes. Drawn for comparison with Fig. 7.2

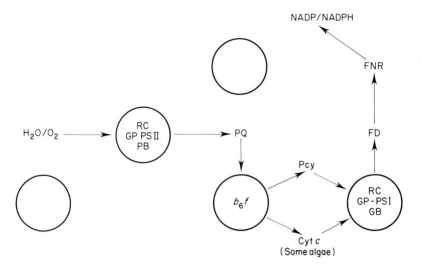

Figure 7.4. Non-cyclic electron transport in thylakoids. RC, photosynthetic reaction centre. PB, purple bacteria. GP PSI, green-plant: photosystem I

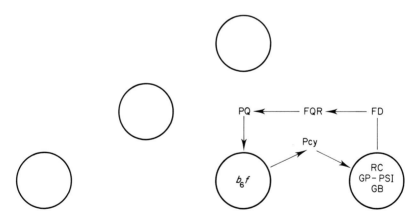

Figure 7.5. Cyclic electron transport around PSI in thylakoids. RC, photosynthetic reaction centre. GB, green bacteria (Chlorobiaceae). GP PSI, green-plant: photosystem I. FD, ferredoxin. FQR, ferredoxin : Q reductase

was generated in the green sulphur bacteria, although it would work in the non-photosynthetic, lithotrophic sulphur-oxidising bacteria, given the proton-translocating properties of cytochrome oxidase (Fig. 7.8).

Figures 7.2 to 7.8 summarise the versatility of the common ground plan of these electron-transport systems. This is chiefly due to their central dependence on the cytochrome *bc* complex, their common segregation of diffusible

148

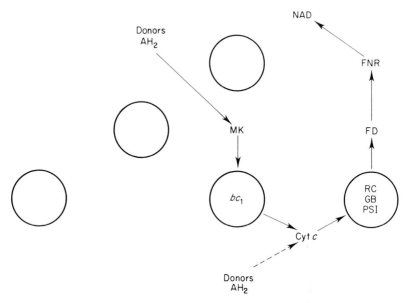

Figure 7.6. Non-cyclic electron transport in Chlorobiaceae. The only justification for including donors acting via menaquinone (MK) is that (i) it provides a coupling mechanism for ATP and (ii) it provides a function for the *bc* complex. RC, photosynthetic reaction centre., GB, green bacteria. FD, ferredoxin. FNR, ferredoxin : NAD(P) reductase

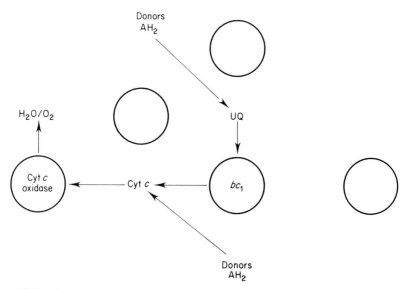

Figure 7.7. Chemosynthesis (chemoautotrophy) in lithotrophic bacteria. The oxidative chain is coupled to ATP by the PMF generated by cytochrome oxidase with or without the involvement of the *bc* complex. NADH may be formed by reverse electron transport (not shown)

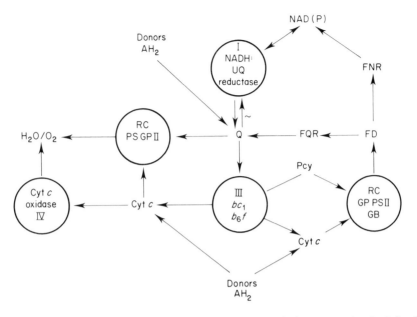

Figure 7.8.   Summary and overview of Figs 7.2–7.7 (abbreviations as previously defined)

substances into three phases, and the common proton-coupled ATPase method of ATP synthesis. There can be little doubt of the evolutionary basis of these common features, but the details remain to be clarified; the appearance of plastocyanin (in place of cytochrome c-553), cytochrome f (in place of cytochrome c_1), the divergence of the two types of photosystem and the appearance of cytochrome oxidase are all hypothetical evolutionary events that may be set in order by mechanistic or genetic studies in the future.

Chapter 8

The Dark Reactions — An Overview

The concept of the two processes in photosynthesis, light and dark reactions, originated with the demonstration by F.F. Blackman that photosynthetic gas exchange was proportional to the light fluence rate up to a certain limit, and the limit was temperature-dependent (Fig. 8.1). By analogy with the photographic process, it was envisaged that light caused an electronic process which did not need activation energy, nor molecular diffusion; one may note that in photography exposure calculations do not include the temperature. Chemical reactions usually do need activation energy, and diffusion of molecules, and so their rates are temperature-dependent, as is the development of a photograph.

Much later, photosynthesis could be seen as divided between membrane-dependent processes and soluble, enzyme-catalysed metabolism. The thylakoid of green plants, and cell membranes or chromatophores of bacteria, generated ATP and NAD(P)H (the light reaction), which were consumed by the surrounding protoplasmic phase (dark reaction). This has been a useful division, and to some extent provides the physical reality behind Blackman's concept. It is necessary to notice, of course, that electron transport in the membrane requires the participation of some diffusible molecules in the adjacent aqueous phases, such as plastocyanin, cytochrome c or ferredoxin, and also that some enzymes are obviously involved in electron transport, notably those that handle the inorganic electron donor (water or sulphur compound), the quinol-binding reductases, FNR and the F-ATPase. The electron transport processes of the light reactions are temperature-dependent except for the primary charge-separating photoreaction.

In green plants and cyanobacteria (and also in the lithotrophic bacteria) the

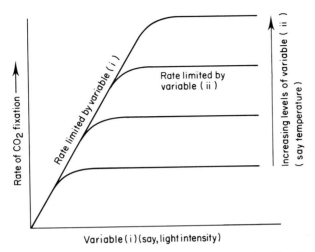

Figure 8.1.   Illustrating the principle that a process such as carbon dioxide fixation may be rate-limited by either of two variables

major dark reaction is the fixation of carbon dioxide by the reductive pentose cycle (Chapter 9). In the green sulphur bacteria carbon dioxide fixation takes place by means of the reductive citrate cycle (subject to some debate, see Chapter 9). Fixation of carbon dioxide is autotrophic nutrition, photoautotrophic in the green organisms above and chemoautotrophic in the lithotrophic bacteria. The reductive pentose cycle is inducible in many species of purple bacteria, but their natural habitat is rich in organic substrates and their normal mode is photoheterotrophic. We use the term 'fermentation reactions' for the anaerobic, ATP-yielding, breakdown of complex cell material into simple excretable organic molecules; photoheterotrophy is simply the reverse of fermentation, taking in the simple molecules and rebuilding them, using the energy of ATP, into the complex cell material. A common environmental carbon source is acetate, and bacteria incorporate it by means of the glyoxylate cycle (Chapter 9).

Bacterial cells are thought of as single (P) compartments, and the products of the light reactions are fed from the outside (the cell membrane). In cells with thylakoids, and chloroplasts, the metabolic reactions of the P phase are fed from the inside. In both cases there are minor photosynthetic processes that make use of reduced ferredoxin and ATP for regulation of enzymic activity, and the fixation of nitrogen (from N_2 or NO_3^-) and sulphur (from SO_4^{2-}).

Compared with prokaryotes, the eukaryote cell has a greater number of compartments. For our purposes, the arrangement can be seen as concentric, with the thylakoid representing the innermost compartment, and the stroma and envelope forming successive surrounding zones or layers. The cytoplasm surrounds the chloroplast with a layer containing other organelles such as peroxysomes and mitochondria. These layers and their organelles act together when carrying out the whole-cell scavenging operation of photorespiration (Chapter 10). The primary photosynthate, triose phosphate, is both stored in the chloroplast by conversion to the polyglucan, starch, and exported to the cytoplasm. In the cytoplasm it is converted via hexose to sucrose (Chapter 9) and stored in the cell vacuole, and in vascular plants sucrose is exported from the photosynthetic tissues as the needs of the plant may dictate.

This expanding concentricity (Fig. 8.2) is very clearly seen in the Kranz anatomy shown by many C4 plants which operate the carbon dioxide pumping system that provides an answer to the photorespiration problem (Chapter 11).

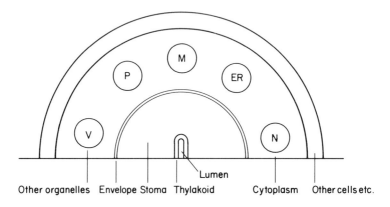

Figure 8.2. Illustrating the concept of concentricity in photosynthesis. The thylakoid is placed at the centre and is surrounded by other regions at increasing distances. V, vacuole; P, peroxisome; M, mitochondrion; ER, endoplasmic reticulum; N, nucleus

In general these patterns of activity are generated by the thylakoid and are transmitted outwards. Influences in the reverse direction are harder to find. One such centripetal effect is the fluorescence induction curve in leaves, in which an inflexion appears after the peak value which is ascribed to the onset of carbon dioxide fixation, although this effect is of more importance to the researcher than to the plant. More recently enzymes in the stroma have been shown to control the phosphorylation of thylakoid membrane proteins, providing feedback control of the light reactions, as described (for the LHCII antenna) in Chapter 3. Other influences are on a longer timescale and include the light-regulated synthesis of membrane proteins in both the stroma of chloroplasts and in the nucleus–cytoplasm system (Chapter 12). The characteristic lipids of the thylakoid membrane are synthesised in the chloroplast envelope (Chapter 1) and transferred inwards by an unknown mechanism. Most of the chloroplast stroma proteins are nuclear coded, and means are required both to convey them across the envelope into the stroma, where they appear to regulate the production in the chloroplasts of the co-subunits of those complexes which have a multiple origin, such as PSI, the *bf* particle and the ATPase, and also the 'most abundant protein', RuBisCO. Chlorophyll, protochlorophyll and phytochrome form a closely interrelated set of pigments for the purpose of regulating the growth, form and movement of photosynthetic organisms, with the obvious function of ensuring that the light availability for the whole organism is optimised. At this arbitrary point of photobiology, however, this biochemical treatment must stop.

Further Reading

The interrelationships of organelles in the cell are usefully described in:

Reid, R.A. & Leech, R.M. (1980) *Biochemistry and Structure of Cell Organelles*, Tertiary Level Biology series, Blackie, Glasgow.

Chapter 9

Fixation of Carbon Dioxide: The Reductive Pentose Cycle

9.1 Description of the Cycle

The reductive pentose cycle is by far the most important process by which carbon dioxide is fixed into organic carbon in the earth's biosphere. It provides a pathway in which carbon dioxide is fixed by reaction with a pentose (ribulose bisphosphate, RuBP), forming two molecules of 3-phosphoglyceric acid (PGA), which are reduced to the level of carbohydrate by ATP and NAD(P)H. The enzymes that carry out the cycle are located in the chloroplast stroma, or the cytosol of prokaryotes, and the light-dependent electron transport process in the membranes provides ATP, and (directly or indirectly) NAD(P)H. PGA is reduced to the level of triose phosphate, in which form it can leave the chloroplast to form the bulk product sucrose, or starch within the chloroplast via hexose phosphate. The substrate for carbon dioxide fixation, RuBP, must be regenerated, and this is achieved by monosaccharide phosphate interconversion reactions similar in effect to the oxidative pentose phosphate pathway in animal tissue.

The pathway represented as a metabolic map in Fig. 9.1 represents the conclusions (175) of a long research project carried out at Berkeley, California, in the laboratory of M. Calvin, for which the Nobel Prize was awarded. The cycle is commonly referred to as the Calvin or Calvin–Benson cycle, or the C3 cycle to contrast with another process known as C4 (Chapter 11).

Figure 9.1 sets out the reductive pentose cycle to show the balance of metabolites as carbon dioxide is fixed to triose phosphate. For each carbon dioxide molecule, two molecules of reduced coenzyme and three ATP are required. The apparent complexity arises from the carbohydrate interconversions that

154

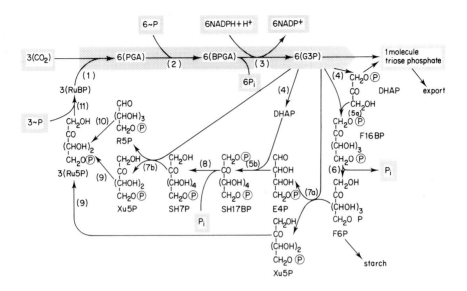

Figure 9.1. The reductive pentose cycle. Drawn to show the formation and export of 1 molecule of triose phosphate (principally in the form of DHAP), from 3 molecules of CO_2. Abbreviations: BPGA, 1,3-bisphospho-D-glycerate; DHAP, dihydroxyacetone phosphate; PGA, 3-phosphoglycerate; G3P, D-glyceraldehyde 3-phosphate; E4P, D-erythrose 4-phosphate; F1,6BP, D-fructose 1,6-bisphosphate; F6P, fructose 6-phosphate; E4P, D-erythrose 4-phosphate; R5P, D-ribose 5-phosphate; RuBP, D-ribulose 1,5-bisphosphate; Ru5P, ribulose 5-phosphate; SH1,7BP, D-sedoheptulose 1,7-bisphosphate; SH7P, sedoheptulose 7-phosphate; Xu5P, D-xylulose 5-phosphate. Enzymes: 1, RuBisCO (EC 4.1.1.39); 2, PGA kinase (EC 2.7.2.3); 3, glyceraldehyde 3-phosphate dehydrogenase (NADP-linked) (EC 1.2.1.13); 4, triose phosphate isomerase (EC 5.3.1.1); 5a, 5b, aldolase (EC 4.2.1.13); 6, FBPase (EC 3.1.3.11); 7a, 7b, transketolase (EC 2.2.1.1); 8, SHBPase; 9, phosphoribulose 3-epimerase (EC 5.3.1.6); 10, phosphoribose isomerase (EC 5.1.3.4); 11, phosphoribulokinase (EC 2.7.1.19)

regenerate the starting material, RuBP; reference to a general text of biochemistry will show the similarity to the oxidative pentose (phosphate) pathway of, say, adipose tissue, the principal enzyme difference being the presence of transaldolase in the oxidative pathway, at the point of formation of sedoheptulose-7-phosphate (SH7P); in the absence of transaldolase, the reductive pathway forms sedoheptulose bisphosphate by means of aldolase, and a bisphosphatase is required. Nevertheless, if the reductive cycle is written out and balanced with transaldolase and compared with the actual state of affairs, it will be seen that the natural pathway is in fact adapted for the minimum number of steps.

The product shown is triose phosphate; that is, the equilibrium mixture of 3-phosphoglyceraldehyde and dihydroxyacetone phosphate (DHAP). In this form (in which DHAP is in the majority) the fixed carbon (photosynthate) is exported to the cytoplasm. The additional processes of starch or sucrose formation increase the requirement for ATP.

9.2 Occurrence in Photo- and Chemoautotrophic Nutrition

The cycle provides the only means of incorporating carbon dioxide into cell material in eukaryotes, and in all prokaryotes except the green sulphur bacteria, in which an alternative exists. Although other carbon dioxide fixation reactions exist in all cells, including those of mammals, such reactions (for example the conversion of pyruvate to OAA) do not lead to any overall fixation, but are rather temporary steps in which the added CO_2 is a leaving group to drive a subsequent reaction.

The reduced coenzyme and ATP required by the cycle can be supplied either by photosynthetic electron transport, or by chemosynthetic reactions in the lithotrophic bacteria. Thus *Thiocapsa*, nominally an anaerobic purple photosynthetic bacterium, can function as an aerobic sulphur-oxidising chemotroph; the reductive pentose cycle is the same in each case. In general, purple photosynthetic bacteria can be adapted to oxidise hydrogen aerobically and fix carbon dioxide by means of the cycle.

9.3 History of the Cycle

9.3.1 The use of *Chlorella*, carbon-14 and paper chromatography

The researches that led to the discovery of the cycle took place in Calvin's laboratory in the 1940s, and have been described many times (see for example the review of Bassham (176)). The project relied on two crucial methods. The first was the use of the radioisotope ^{14}C in the form of $^{14}CO_2$. The alga *Chlorella* in suspension was illuminated in the presence of unlabelled CO_2 and continuously sampled. $^{14}CO_2$ was added to the sampling stream which flowed through a transparent, illuminated tube so that the isotope was incorporated into the cells for a predetermined period, which could be as short as 9–11 seconds. The stream then flowed into hot alcohol, which stopped the reaction. When sufficient material had accumulated, the metabolic intermediates were analysed. The actual quantities of the metabolites did not change appreciably, but the radiolabel was expected to be distributed among a number of intermediates that was proportional to the time of illumination. Extrapolation back to zero time would identify the earliest product of carbon dioxide fixation. In fact the analysis was impracticable until the second, crucial method became available: paper chromatography, invented by A.J.P. Martin's group (177).

The mixture of metabolites to be analysed was placed at the origin of a sheet of chromatography paper (filter-paper) and dried. A solvent was allowed to pass through the paper from one edge to the opposite edge, by capillary action or gravity, and the different materials in the mixture were spread out in spots along the line of movement of the solvent, according to their different mobilities. The paper was dried, and a second different solvent (in which the mobilities of the spots were different) applied in a direction at right-angles to the first. Each substance ideally migrated to form its own distinct spot that could be visualised by means of stains or colour reactions. Paper chromatography has nowadays been largely superseded by thin-layer techniques.

The spots on the paper were identified by reference to standards, and the degree of radiolabelling in each spot measured by means of a radioautograph, where the chromatogram was placed against photographic (X-ray) film, kept in the dark for a sufficient period of time, and subsequently developed (see Fig. 9.2). PGA was identified as the first compound to be formed from labelled carbon dioxide, and chemical analysis showed that the label initially appeared only in the carboxyl group. Later the label appeared in all three carbon atoms of PGA, proving that its precursor was regenerated from PGA: in other words, that a cycle existed. The reactions of the cycle were brilliantly deduced, although it was criticised at the time on the grounds that not all the enzymes had been located in the chloroplast. The first working model used a mixture of enzymes from various sources, including animals.

9.3.2 Problems of RuBisCO kinetics—maximum rate and control

The enzyme catalysing the fixation reaction is known as RuBisCO (ribulose bisphosphate carboxylase oxygenase, abbreviated also to RuBPCase, EC 4.1.1.39) because it has two competing activities, one with the substrate carbon dioxide that is part of the pentose cycle, and a second reaction with oxygen that forms glycollate and is part of the photorespiration pathway (Chapter 11). Earlier names for the enzyme were carboxydismutase and fraction-I protein. RuBisCO is a large enzyme with a complex quaternary structure. It is usually made up of eight large subunits (A subunits) and eight small subunits (B), so that the commonest forms are represented as A_8B_8. The A subunit has a mass of 55 kDa, and the B subunit a mass of 12–16 kDa, so that the total structure is of the order of 550 kDa. The large subunit is coded and synthesised in the chloroplast, and the small subunit is coded in the nucleus, and synthesised as a precursor polypeptide in the cytoplasm (see Chapter 12).

Although RuBisCO is the most abundant protein in the world, accounting for almost half the protein of most leaves, as an enzyme it has a low turnover rate. For some years after its discovery, the maximum rate attainable by the isolated enzyme was insufficient to account for the observed rates of photosynthesis. Studies revealed that this enzyme has in fact a range of activating systems that is unique in variety and scale.

The affinity of RuBisCO for carbon dioxide

The Michaelis constant (K_m) for carbon dioxide in preparations of RuBisCO obtained up to 1966 was considerably higher than the concentration of carbon

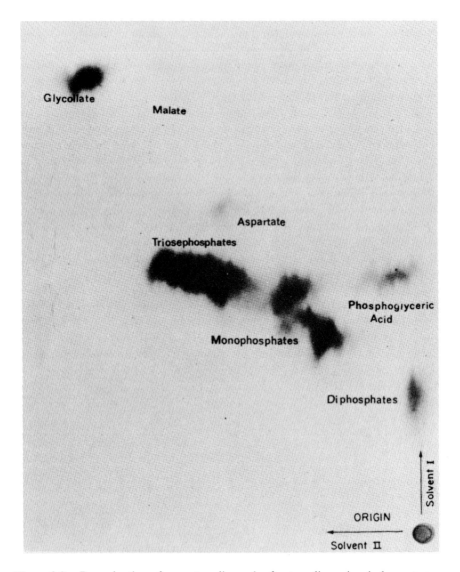

Figure 9.2.   Reproduction of an autoradiograph of a two-dimensional chromatogram on paper of the products of photosynthesis in pea chloroplasts using $^{14}CO_2$. The dark spot due to radioactive triose phosphate indicates that the principal pathway was the reductive pentose cycle, other intermediates of which can be seen as paler spots. The occurrence of glycollate is common and is discussed in Chapter 10. Courtesy of Professor D.A. Walker

dioxide attainable from the air. This meant that the activity to be expected in the chloroplast would be too low to support the observed rate of carbon dioxide fixation, and hence doubt was cast on the validity of the cycle. It was then shown by Jensen and Bassham (178) that the enzyme could be isolated in a form with a much lower K_m if the protein concentration was kept high. It later appeared that the two forms of the enzyme (high and low K_m) can be interconverted. Incubation with carbon dioxide or hydrogen bicarbonate and magnesium ions (10 mM) (in the absence of RuBP) lowers the K_m, and conversely incubation with RuBP raises it. For practical purposes the enzyme is considered to be 'activated' when in the low-K_m state. In the activation process a molecule of carbon dioxide is bound at the regulatory site (lysine 201 on the A subunit) and is not directly involved in the catalytic process. The changes in K_m take minutes to complete, and it may be suppposed that there is an accompanying readjustment of the relationship of the subunits (eight each of the A and B subunits in green plants), probably following the formation of magnesium carbamylates of amino groups on the protein:

$$E\text{-}NH_2 + CO_2 + Mg^{2+} > E\text{-}NH\text{-}COO^- \ Mg^{2+}$$

A similar change in quaternary structure follows carbamylation in haemoglobin ($\alpha_2 \ \beta_2$), but very much faster. Reversible conformational changes have also been identified as the basis of the cold inactivation of RuBisCO.

RuBisCO isolated from one bacterium (*Rhodospirillum rubrum*) does not contain B subunits (it is A_2), and it has been shown that catalytic activity remains in the A_8B_8 types after removal of the B subunits, but the activation phenomena described above are lost. Therefore the B subunit is believed to approach close to the catalytic site in such a way as to bring the lysine-(A)-201 on which the magnesium carbamate is formed, into association with the catalytic centre (see the review by Akazawa (179)).

It has been shown that RuBisCO exists in two fractions, the major one being a phosphorylated form. There is no evidence that the phosphorylation is related to gain or loss of activity, and it may be associated with the mechanism for transporting the small subunit into the chloroplast, and with the system for assembling the relatively insoluble large subunit into the octameric arrangement with the small subunit, which takes place by means of a RuBisCO-A-binding protein.

RuBisCO is also activated in the chloroplast stroma both by the rise in pH from 7 to 8.5 and by the rise in the concentration of magnesium ions from less than 1 to some 5 mM, which occur when the thylakoid is illuminated and sets up the electrochemical gradient. The full explanation may lie in the magnesium–carbon dioxide activation mechanism proceeding faster under these conditions, since the effect on the catalytic constants themselves is not sufficient to explain the great change in rate.

Recently protein activators have been identified. A gene product (i.e. a protein) is required for activation of RuBisCO in vivo (180); in its absence the plant needs very high levels of carbon dioxide to survive. This protein activator is thought to contain two polypeptide chains. It may provide the rationale for the earlier report of the need to maintain high protein concentrations during purification of RuBisCO (178).

A non-protein inhibitor which is probably a metabolic analogue of RuBP has been shown to accumulate during darkness, and to be destroyed within minutes of the onset of illumination. It has been called the 'pre-dawn inhibitor', and after purification it was identified as 2′-carboxyarabinitol-1-phosphate (181). There is a structural similarity with the hypothetical enzyme-bound intermediate formed during the carboxylase reaction: 2-carboxy-3-ketoribitol-1,5-bisphosphate. It was noted (182) that the related substance carboxyarabinitol-1,5-bisphosphate, which does not occur naturally, is an irreversible inhibitor. It may not be completely sufficient to describe the 1-phosphate as a inhibitor, however, since the compound binds to the activated form of RuBisCO and maintains it in the active state. It was pointed out that the high affinity meant that it was effective in a low concentration, equivalent to that of RuBisCO, and hence catalytic activity was able to start immediately with the onset of the removal of the dawn inhibitor. Previous hypotheses concerning pre-dawn inhibitors were based on 6-phosphogluconate which was required to exist in relatively high concentrations, and could not be readily disposed of.

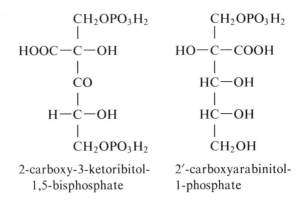

2-carboxy-3-ketoribitol-
1,5-bisphosphate

2′-carboxyarabinitol-
1-phosphate

The activation of RuBisCO affects both the oxygenase and the carboxylase activities equally; there is no proven means of differentially inhibiting the oxygenase reaction in spite of much research propelled by the obvious impact such an inhibitor would have on agriculture (see Chapter 11).

One major net effect of all the regulatory processes described above is to ensure that there is no measurable activity of RuBisCO under normal conditions in the dark. This is essential, because the reductive pentose cycle is an

ATP-consuming system, and in the absence of ATP from the light reactions of photosynthesis, the chloroplast will draw in ATP and reduced coenzymes from the cytoplasm, where they are produced by oxidative reactions. Unless the reductive pentose cycle can be turned off, there will in effect be an attempt at a perpetual motion machine, synthesising carbohydrate in chloroplasts and oxidising it in the cytoplasm. This would result in a rapid net drain of ATP and hence of fixed carbon. The oxygenase activity of RuBisCO would pose an even worse threat.

9.3.3 The Gibbs effect

The cycle as drawn in Fig. 9.1 shows that hexose is formed from trioses, and the trioses are interconvertible. In fact the aldolase and isomerase enzymes are present in high activity and would be expected to be close to equilibrium. It was therefore surprising when Gibbs (183) showed that the labelling pattern of hexose was not consistent with its supposed formation from an equilibrium mixture of aldo- and ketotriose phosphates.

The solution to the paradox (illustrated in Table 9.1) appears to lie in the combined effect of (i) the much smaller equilibrium concentration of glyceraldehyde-3-phosphate (GAP) in relation to that of DHAP, and (ii) the lower rate of transport of the former across the chloroplast envelope. The latter is therefore expected to be more rapidly diluted by non-labelled material from outside the chloroplast, and hence the radiolabelling will be less for the C1 to C3 of hexose (which are derived from DHAP) with respect to the corresponding carbons C6 to C4 (derived from GAP).

Table 9.1   The Gibbs effect (from Gibbs (183))

Time (sec)	Carbon atoms in polysaccharide glucose					
	1	2	3	4	5	6
5	5.0	5.4	73	100	0.8	0.8
30	20	17	86	100	14	18

The figures show the specific activity of the carbon atoms of glucose from cells of *Chlorella pyrenoidosa* relative to C-4.

9.4 The Other Enzymes of the Cycle

The kinases phosphoglycerate kinase (PGK) and phosphoribulokinase (PRuK) consume the three molecules of ATP accompanying the fixation of each molecule of carbon dioxide.

PGK in the chloroplast is believed to be a different gene product from the enzyme with the same activity in the cytoplasm, but the properties of the two are very similar. This enzyme catalyses a reversible reaction, unlike most other kinases, and in both localities the reversibility is required by the hypothesis of shuttle mechanisms for transport of ATP across membranes (Chapter 12). The reversibility stems from the fact that the phosphate forms a mixed anhydride at C1, so that the Gibbs energy of hydrolysis is comparable with that of ATP. The regulation of the enzyme is simple in that the ATP-consuming (glucogenesis) reaction is competitively inhibited by ADP (with respect to ATP). This may not be relevant, however, since it has been shown (184) that the rate of carbon dioxide fixation did not require particularly high values of the 'driving force' (F_A) calculated by:

$$F_A = [\text{ATP}][\text{NADPH}]/[\text{ADP}][\text{P}_i][\text{NADP}^+]$$

PRuK on the other hand shows a complicated pattern of regulation, particularly in bacteria, and its activators (ATP and dithiothreitol) appear to dissociate a tetrameric form to the monomer, providing potentially a means for the regulation of the cycle by light. The behaviour is similar to that of glyceraldehyde-3-phosphate dehydrogenase (see below).

Glyceraldehyde-3-phosphate dehydrogenase (GAPDH) in the cytoplasm is specific for NAD^+ as coenzyme, but the chloroplast contains both NADP^+–GAPDH and NAD^+–GAPDH activities and these may be the same enzyme. The bacterial enzyme is specific for NAD^+. The chloroplast enzyme resembles PRuK in size and being a tetramer, dissociating in the presence of an activator such as ATP, dithiothreitol, NADP^+ or NADPH. This may explain the two- to six-fold activation that takes place in illuminated leaves. The requirement for reducing agent may be met in chloroplasts by the thioredoxin system. Observations by Bradbeer et al (185) suggested that the association–dissociation behaviour of GAPDH is absent in C4 plants such as maize, implying a lesser need for regulation, and conversely, that regulation is significant in C3 photosynthesis.

Triose phosphate isomerase in the chloroplast is virtually identical to the cytoplasmic enzyme. It is inhibited by phosphoenolpyruvate (PEP), RuBP and phosphoglycollate. The cytoplasmic enzyme is present in sufficient quantity in many tissues to ensure that there is virtually always an equilibrium between GAP and DHAP. However, the Gibbs effect (differential labelling of C3 and C4 in photosynthetic hexose) has been interpreted as indicating differential mixing of cytoplasmic and stromal triose phosphates across the envelope, and for this to cause the effect, DHAP and GAP would need to be not in equilibrium.

Aldolase reactions appear at two points in the cycle, catalysing the formation of fructose bisphosphate (FBP) and sedoheptulose bisphosphate (SHBP) from DHAP with GAP and erythrose phosphate respectively. Both activities co-purify and they are believed to be the same enzyme.

Transketolase acts on fructose and sedoheptulose phosphates, removing the carbon atoms 1 and 2 as a ketol group attached to the coenzyme thiamine pyrophosphate (Fig. 9.3), and transferring them to C1 of GAP, forming Xu5P. The residue of the F6P is erythrose-4-phosphate (E4P), the substrate for SHBP aldolase, and the residue from the SH7P is R5P, to be converted to Ru5P.

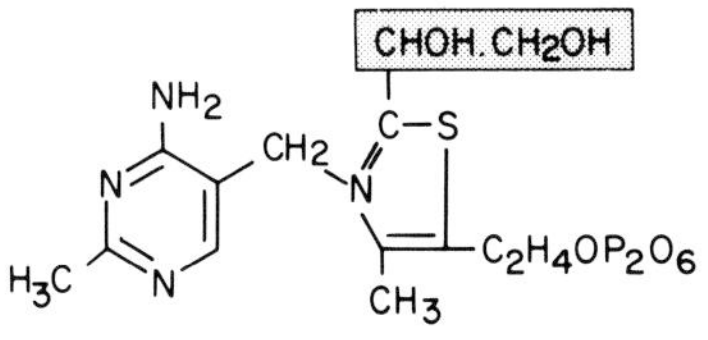

Figure 9.3. The intermediate in the transketolase reaction. The shaded group is the ketol (C_2) fragment attached to the coenzyme thiamine pyrophosphate

FBP and SHBP phosphatases are different enzymes, producing F6P and SH7P respectively and liberating phosphate. Both are activated in light, by means of the thioredoxin system. FBPase is, in addition, strongly pH-dependent, and is regarded as the major controlling enzyme of the cycle (186). (Controlling in this context implies making substantial changes in the overall rate of the cycle to conform with external circumstances, as opposed to 'fine-tuning', whereby the rates of individual steps are conformed with the overall rate, or 'on-off switching', which is the means by which the cycle is completely inactivated in the dark.) SHBPase was considered to present a problem in that its activity was too low to reconcile with the rate of carbon dioxide fixation, but it appears to be inhibited by its product, SH7P.

Phosphoribulose 3-epimerase (EC 5.1.3.1.) converts Xu5P to Ru5P, and phosphoribose isomerase (EC 5.3.1.6) converts R5P to Ru5P. These are both equilibrium reactions. Ru5P is the substrate for PRuK, regenerating RuBP, the substrate for RuBisCO.

9.5 Relationships of the Pentose Cycle

Intact chloroplasts evolve oxygen when illuminated in the presence of carbon dioxide and phosphate, and rates of CO_2 fixation of 50–250 μmol (mg Chl)$^{-1}$h^{-1} are readily observed. Such chloroplasts have been termed Type A by Hall (248), and do not perform the Hill reaction because ferricyanide does not penetrate the envelope. As the envelope becomes leaky the rate of fixation of carbon dioxide decreases sharply (Type B), and the ferricyanide-dependent Hill reaction appears. A comparative discussion of methods for preparing chloroplasts is given by Jensen (249).

A lag is usually observed, termed photosynthetic induction. It has been studied by Walker using a reconstituted system containing the stroma enzymes but free from membrane compartmentation. The lag can be eliminated by adjustment of the concentrations of enzymes and substrates. Induction in isolated chloroplasts is explained partly by the need to activate enzymes, and partly by the adjustment and build-up of the concentration of intermediates, among which PGA and Ru5P are considered most crucial. If phosphate is present in the medium at a higher concentration than inside the chloroplast, the transporter in the envelope will exchange it for PGA, DHAP, etc. (see Chapter 12), diminishing the total metabolite pool and producing the effect of an apparent inhibition of photosynthesis by phosphate.

The cycle is able to adjust to the loss of any particular metabolic intermediate. The proportions of reactants shown in Fig. 9.1 are calculated to account for the drain of triose or hexose phosphates, but variations can be recalculated to allow any other intermediate to be produced instead; the cycle can be said to be auto-anaplerotic, meaning that it can restore the levels of its intermediates.

The thioredoxin system

Thioredoxin is a Fe_4S_4 protein with two different peptide chains, about 13 kDa in mass. The protein reacts with iodoacetate in the reduced form only, and one of the subunits contains a disulphide group that is reduced via ferredoxin by chloroplast thylakoids in light (187). It is supposed that thioredoxin is the natural source of reductant for enzymes that in vitro are activated by thiols such as dithiothreitol. Thioredoxin could provide a mechanism for the control of enzyme activity, in that in the dark when there is no reductant the SH-enzymes are allowed to become oxidised. However, at the present time the only enzyme in which reduction of S–S to $(SH)_2$ has actually been demonstrated is FBPase, and it is difficult to distinguish a possible regulatory purpose from a maintenance system, preventing the continuous oxidation of SH groups by the high concentration of oxygen.

The biosynthesis of starch

Most green plants synthesise starch both in the chloroplasts and in starch grains (amyloplasts, related to chloroplasts) in other tissues, especially in reserve organs.

The pathway of chloroplast starch synthesis appears in Fig. 9.4, which shows that the crucial step is the formation of ADP-glucose from G1P and ATP. This is in contrast to (e.g.) glycogen formation in animal tissues where UDP-glucose is the intermediate. The older hypothesis that starch was formed by the reversal of phosphorylase

$$\text{starch} + P_i \longrightarrow \text{glucose-1-phosphate}$$

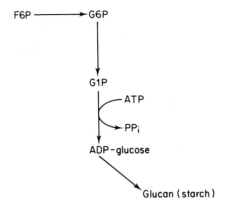

Figure 9.4. Formation of starch (see Fig. 9.1)

was ruled out in mammals because the controls on phosphorylase were more appropriate to its role in catabolism, and because phosphorylase deficiency caused glycogen accumulation. In the chloroplast the evidence is not so clear cut (see the review by Preiss and Levi (188)). A minor alternative to ADP-glucose is UDP-glucose (the form used for mammalian glycogen formation), and UDP-glucose is possibly an available substrate for the synthesis of starch in reserve granules in non-photosynthetic tissue.

The formation of ADP-glucose is allosterically activated by PGA and inhibited by phosphate, and this may provide a control mechanism, such that starch synthesis does not begin until the export of triose phosphate to the cytoplasm for sucrose synthesis has reached saturation (see Chapter 12).

9.6 Carbon Metabolism in the Phototrophic Bacteria

9.6.1 Photoautotrophy: the reductive pentose cycle

Purple bacteria are able to fix carbon dioxide, after adaptation by the synthesis of induced enzymes. They possess the enzyme RuBisCO, often packaged into polyhedral bodies known as carboxysomes, and perform the reductive pentose cycle. The countersuggestion that the purple sulphur bacteria such as *Chromatium* might share the reductive citrate cycle (see below) with the green sulphur bacteria (both groups are strict anaerobes) was eliminated by the demonstration (189) that the isotope discrimination (between $^{11}C/^{12}C$) was the same as in the C3 plants in both groups of purple bacteria, but was different in the green sulphur bacteria such as *Chlorobium*.

The reductive pentose cycle is unquestionably the dominant pathway of carbon dioxide fixation in the purple bacteria. The green non-sulphur bacterium

Chloroflexus has a metabolic pattern very similar to that of purple non-sulphur bacteria.

The cycle is also established for the chemiautotrophic bacteria, of all the various groups: for example, *Thiobacillus thiooxydans* (sulphur-oxidising), *T. denitrificans* (nitrate-reducing), *T. ferrooxidans* (Fe^{2+}-oxidising) and *Hydrogenomonas* (hydrogen- and oxygen-reacting).

9.6.2 Photoautotrophy: the reductive citrate cycle

The reader will be familiar with the (oxidative) citrate cycle (Fig. 9.5) (tricarboxylic acid cycle), associated with the Nobel Laureate Sir Hans Krebs, and central to the metabolism of most aerobic, chemotrophic cells as well as in plant and purple bacteria cells in the dark. There are many biological functions performed by the cycle, of which we may mention four: production of carbon dioxide, production of ATP at substrate level, production of reduced coenzyme for production of ATP at respiratory chain level, and the interconversion of carbon skeletons, especially for amino acid synthesis.

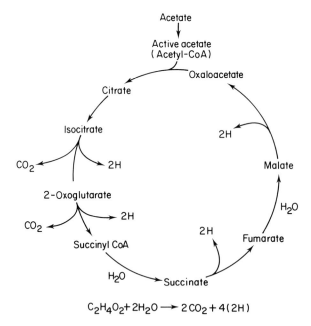

$$C_2H_4O_2 + 2H_2O \longrightarrow 2CO_2 + 4(2H)$$

Figure 9.5. The oxidative citrate cycle (tricarboxylic acid cycle, Krebs cycle). The main pathway for the energy-yielding oxidation of fat and carbohydrate in most aerobic organisms. The (2H) units shown enter the electron transport chain (see Fig. 7.2). The cycle is fed with 'active acetate' (acetyl CoA). Acetyl CoA may be derived from various sources, but carbohydrate is shown here in order to enable comparisons to be made with the various pathways of CO_2 fixation in Figs 9.1, 9.6 and 9.7

The reductive citrate cycle was proposed by Evans et al (190) on the basis of enzymes discovered in green and other bacteria, in the laboratory of Arnon. The reactions (Fig. 9.6) are catalysed by the same enzymes as in the oxidative version, with the exception of four. These are 2-oxoglutarate synthase, pyruvate synthase, pyruvate phosphate dikinase and possibly the citrate cleavage enzyme (if citrate lyase is not present; the difference between the two enzymes is the requirement for ATP in the cleavage enzyme). With the exception of the splitting of citrate, which may be enough of an equilibrium reaction without the need for ATP, all the other steps represent adaptations which effectively reverse reactions which in the oxidative cycle are regarded as irreversible.

The oxoacid synthases reverse the corresponding dehydrogenase reactions because, whereas the dehydrogenases reduce nicotinamide coenzymes ($E_{m,7}$ = -340 mV), the synthases are driven by reduced ferredoxin ($E_{m,7}$ values for bacterial ferredoxins are in the range -410 to -480 mV but no value for Chlorobiaceae ferredoxin is available), and the additional reducing power of the ferredoxin makes the difference. The pyruvate phosphate dikinase is the same as in the C4 photosynthesis pathway (Chapter 11), and relies on the immediate breakdown of pyrophosphate to achieve the synthesis of PEP. The other product, AMP, requires a further ATP to form two molecules of ADP by the adenylate kinase reaction.

Other enzymes of the citrate cycle catalyse equilibrium reactions, and there is no conceptual difficulty in their proceeding in the direction of synthesis.

The cycle is energy-efficient compared to the reductive pentose cycle. If the requirements of (2H) and ATP are totalled for the formation of triose phosphate from three carbon dioxide molecules (treating the pyruvate phosphate dikinase reaction as equivalent to the conversion of two molecules of ATP to two ADP plus one phosphate), in the reductive pentose cycle of Fig. 9.1, and the reductive citrate cycle of Fig. 9.6, the citrate cycle has a marked advantage, requiring four ATP compared to nine in the pentose cycle.

$$3CO_2 + 6(2H) + 9ATP \rightarrow C_3H_6O_3\text{-}PO_3H_2 + 9ADP$$
$$+ 8P_i + 3H_2O \dots \text{(RPC)}$$

$$3CO_2 + 6(2H) + 4ATP \rightarrow C_3H_6O_3\text{-}PO_3H_2 + 4ADP$$
$$+ 3P_i + 3H_2O \dots \text{(RCC)}$$

The reductive citrate cycle is therefore an appropriate pathway for the green sulphur bacteria which live at very low light levels. Indeed it is not at all obvious why the citrate cycle, rather than the pentose cycle, should not have been adopted by the green plants.

It has to be noted that the enzyme citrate lyase is not so well documented in *Chlorobium* cells as the other three special enzymes, and therefore there

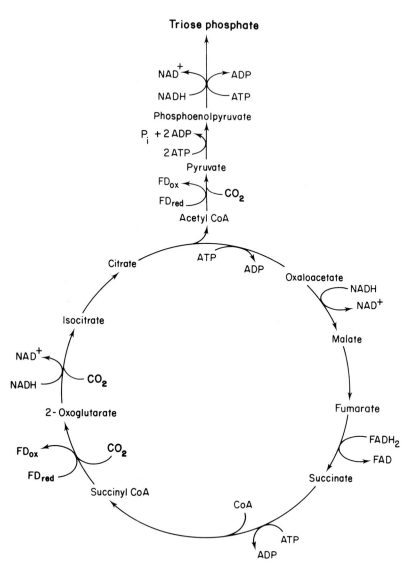

Figure 9.6.   The reductive citrate cycle. The irreversible steps of the oxidative cycle (Fig. 9.5), the oxidative decarboxylation of the 2-oxoacids, are reversed in the green sulphur bacteria by means of the reductant ferredoxin, which has a lower redox potential than NAD^+

is room for debate as to the status of the citrate cycle. Isotope discrimination evidence (see Section 9.6.1) is the chief experimental support for the operation of the citrate cycle, in the green sulphur bacteria only.

9.6.3 Photoheterotrophy: the glyoxylate pathway

A common habit of many purple non-sulphur bacteria is the photosynthetic incorporation of organic materials such as acetate, for which ATP and NADH are required. (In some cases the reductive pentose cycle can be induced.) Acetate can not serve as a feedstock for biosynthesis in many cells because it can only enter the oxidative citrate cycle, in which two carbon atoms are lost as carbon dioxide between isocitrate and succinyl CoA (see Fig. 9.5). The glyoxylate cycle provides a bypass for the two decarboxylations, by means of the enzymes isocitrate lyase (isocitritase) and malate synthase. Isocitrate lyase splits isocitrate into succinate and glyoxylate ($CHO.COOH$), and malate synthase combines glyoxyate with acetyl CoA to form malate in a reaction exactly analogous to that catalysed by citrate synthase. The resulting modified citrate cycle is able to incorporate carbon from acetate into cell material; in Fig. 9.7, the cycle is shown synthesising the same intermediates as the reductive citrate cycle (Fig. 9.6), in this case

$$3CH_3COOH + 5ATP + H_2O \rightarrow C_3H_5O_3\text{-}PO_3H_2 + CO_2 + 4ADP + 4(2H)$$

acetate   triose phosphate

ATP is required in the cycle, as shown, for the conversion of acetate to acetyl CoA (acetyl CoA synthase).

The glyoxylate cycle is capable of adjustment to a variety of substrates, and the overall redox balance of the carbon nutrient supplied can be adjusted to the balance of the cell material as a whole, by the cell evolving either carbon dioxide (as shown in the equation above) or hydrogen (see below).

9.6.4 The phosphoroclastic reaction

2-Oxoacids such as pyruvate can be oxidatively decarboxylated to acetyl phosphate, reducing ferredoxin, which is mainly used for nitrogen fixation. It is a property of the nitrogenase system that some hydrogen is always produced, as a wasteful side-reaction, and in the absence of nitrogen the production of hydrogen becomes quantitative, so that photosynthetic bacteria are observed to decompose organic acids into carbon dioxide and hydrogen.

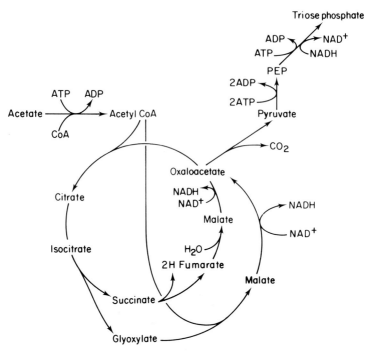

Figure 9.7. The glyoxylate pathway. The presence of the enzyme iso-citrate lyase (isocitritase) allows the CO_2-producing stages of the oxidative citrate cycle (Fig. 9.5) to be avoided. Hence acetate can be converted to CO_2, triose phosphate and 2(2H). This is a common activity in photoheterotrophic bacteria. ATP is required (from the light reactions) but no reductant; the (2H) units are lost as H_2

Chapter 10

Photorespiration

10.1 Definition: Measurement of Carbon Dioxide Exchanges

Photorespiration is the light-dependent increase in the rate of carbon dioxide production, observed in photosynthetic cells, that is associated with the production of glycollate. Its existence is not immediately obvious because the simultaneous consumption of carbon dioxide in the reductive pentose cycle is considerably greater, and masks the carbon dioxide output. However, when the light is turned off the production of carbon dioxide continues for several seconds, as a post-illumination carbon dioxide burst, observed and studied by Decker (191). A more quantitative method is to measure the steady-state uptake of $^{12}CO_2$ and then apply a pulse of $^{14}CO_2$. The net uptake of carbon dioxide is measured by infrared analysis of the gas stream passing through a chamber containing a leaf, and the uptake of the radioisotope measures the gross photosynthesis; the rate of photorespiration can be calculated by subtraction. It became apparent that the newly fixed label reappeared in a few seconds as carbon dioxide, and this showed that the carbon dioxide produced in photorespiration came preferentially from newly fixed carbon and not from the cell pool of metabolites (see the review of Zelitch (192)).

Photorespiration is to be distinguished, in theory at least, from an increase in the general level of metabolic activity and (mitochondrial) respiration that occurs in algae, triggered by blue light (one of the 'blue light effects'). This effect is related to an increase in protein synthesis (193).

Photorespiration is also to be distinguished from any continuing mitochondrial respiration (which is usually found to be suppresssed during photosynthesis in leaves: see Chapter 12).

10.2 Production of Glycollate in *Chlorella* and *Nicotiana*

The earliest paper chromatograms of the products of $^{14}CO_2$ incorporation in *Chlorella* by Calvin's group always showed glycollate spots, which were (correctly) ignored in the formulation of the reductive pentose cycle. However, it was clear from this work that glycollate production was indeed a major activity of *Chlorella*, and that considerable quantities of it were excreted by the cells into the medium. The production of considerable quantities of glycollate was also observed in the tobacco leaf (*Nicotiana*) in the studies of Zelitch.

The source of much of the glycollate was identified by Bowes et al (194) as RuBisCO acting as an oxygenase on the (first) substrate RuBP. RuBisCO accepts either carbon dioxide or oxygen as the second substrate. (RuBisCO is activated by an enzyme-bound carbon dioxide molecule; the substrate oxygen or carbon dioxide molecule is probably not bound to the enzyme (195).) In the oxygenase reaction, oxygen attacks the bound RuBP which is split to give one molecule of PGA and one molecule of phosphoglycollate (Fig. 10.1). Phosphoglycollate phosphatase (a stroma enzyme) releases glycollic acid and phosphate.

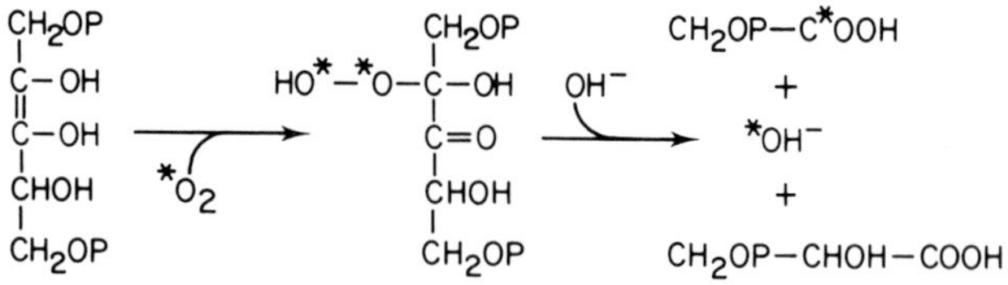

Figure 10.1. The production of glycollate. Attack by O_2 on the enediol form of RuBP. The asterisk indicates the fate of the O atoms of O_2

Other sources of glycollate

It became clear from isotope work with $^{18}O_2$ that although 70–80% at least of glycollate was labelled in the carboxyl position with ^{18}O and therefore came from an oxygenase reaction, the existence of other sources could not be excluded. Possible alternative sources are a non-enzymic oxidation of RuBP by hydrogen peroxide, which is formed in the oxidation of glycollate (see below), or an oxidative attack on the C_2 (ketol) group carried on thiamine pyrophosphate during the reaction catalysed by transketolase (Fig. 9.3).

10.3 The Scavenging Pathway of Photorespiration

The pathway to be described has been given the title 'the oxidative photosynthetic carbon cycle of photorespiration' (195). The only connection with photosynthesis, however, is the common starting point, RuBP. The concept is now widespread that this pathway represents an unavoidable leak in the reductive pentose cycle, and a means of clearing-up the consequences, hence 'scavenging'.

Isolated chloroplasts do not photorespire; the output of carbon dioxide is a whole-cell phenomenon. Glycollate is oxidised by an enzyme containing FMN in the peroxysome fraction of the cell, yielding glyoxylate and hydrogen peroxide:

$$CH_2OH.COOH + O_2 \rightarrow CHO.COOH + H_2O_2$$

This is the second point at which oxygen is consumed in the photorespiratory process. Half a molecule of oxygen is regenerated when the hydrogen peroxide is decomposed by catalase, which is abundant in the peroxysome. At high rates of reaction of the glycollate oxidase, the concentration of peroxide in the cell could rise to levels sufficient to provide for the alternative reactions producing glycollate, suggested above. Catalase also has a peroxidatic activity which is most marked when hydrogen peroxide is continually being produced at low concentration. It might be expected that glyoxylate would be peroxidised by catalase (or even non-enzymically) to yield formate and carbon dioxide, but this does not happen to a significant extent, for reasons which are unclear.

Glyoxylate is transaminated to glycine, also in the peroxysome. The amino donor in principle could be glutamate, alanine or serine, but the scheme as drawn (Fig. 10.2) is at its simplest with equal contributions from serine and glutamate. These amino acids become hydroxypyruvate and oxoglutarate respectively. The glycine passes from the peroxysome to the mitochondrion where two molecules of glycine give rise to one molecule each of serine, carbon dioxide and ammonia, by the well known reactions using the coenzymes tetrahydrofolic acid and NAD^+. This is the point in the cycle where the photorespiratory carbon dioxide is produced, half a molecule per original molecule of glycollate. The ammonia is taken up by oxoglutarate using the coenzyme NADH and regenerating the glutamate.

Returning to the peroxysome, the serine transaminates incoming glyoxylate, becoming hydroxypyruvate which is reduced by NADH to glycerate. Glycerate returns to the chloroplast and is phosphorylated by means of ATP and glycerate kinase to 3-phosphoglycerate.

The overall process represented in Fig. 10.2 can be summarised:

$$2RuBP + 3O_2 + NADH + ATP \rightarrow 3PGA + 2P_i$$
$$+ CO_2 + 2H_2O + NAD^+ + ADP$$

The PGA re-enters the pentose cycle, and regenerates RuBP; one tenth of the carbon of RuBP has been lost as carbon dioxide. If the above equation is reformulated in a cycle in which 10 molecules of RuBP give rise to 15 molecules of PGA, which then regenerate 9 molecules of RuBP, the overall effect is the

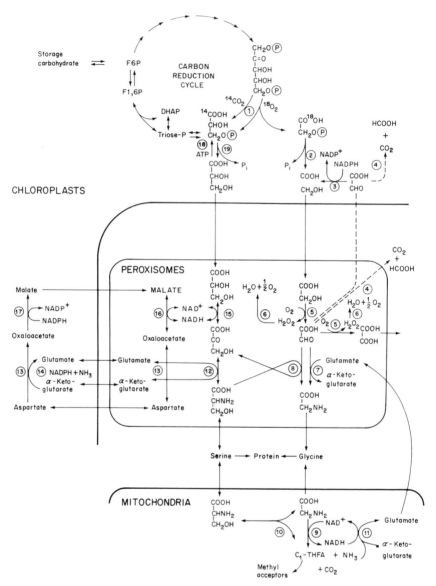

Figure 10.2. The glycolate pathway. The numbers are for the name of the enzyme or reaction: *1*, RuBisCO; *2*, P-glycolate phosphatase; *3*, NADPH glyoxylate reductase; *4*, glyoxylate oxidation by any oxidant; *5*, glycolate oxidase; *6*, catalase; *7*, glutamate–glyoxylate aminotransferase; *8*, serine–glyoxylate aminotransferase; *9*, glycine oxidase; *10*, serine hydroxymethyl transferase; *11*, NADH–glutamate dehydrogenase; *12*, glutamate–hydroxypyruvate aminotransferase; *13*, glutamate oxaloacetate aminotransferase or aspartate aminotransferase; *14*, NADPH–glutamate dehydrogenase; *15*, NADH–hydroxypyruvate reductase or glycerate dehydrogenase; *16*, NAD–malate dehydrogenase; *17*, NADP–malate dehydrogenase; *18*, glycerate kinase; *19*, P-glycerate phosphatase. Adapted from Tolbert (195)

destruction of one molecule of RuBP, and the loss of ATP and reduced coenzymes, as shown:

$$RuBP + 15O_2 + 5NADH + 29ATP + 15NADPH$$
$$\rightarrow 5CO_2 + 25H_2O + 5NAD^+ + 15NADP^+ + 31P_i + 29ADP$$

This may serve to demonstrate how much of an energy drain as well as a carbon drain the photorespiration pathway is.

One turn of the cyclic scheme of Fig. 10.2 accounts for the loss of one carbon from the chloroplast metabolite pool and hence explains the preference for newly fixed carbon dioxide. The dependence on light is due to the need for NADPH and ATP to re-form RuBP, and for the activation of certain enzymes by means of the thioredoxin system (see Chapter 9). When illumination ceases, chloroplast reactions stop within seconds, but glycollate continues to diffuse from the chloroplast, through a specific transporter (196). The post-illumination carbon dioxide burst lasts during the time taken for glycollate to diffuse to the peroxysome.

10.4 Compensation Points in Relation to Photorespiration, and the Warburg Effect

Since oxygen and carbon dioxide compete for the enzyme RuBisCO, photorespiration will be more serious at high levels of the ratio of the partial pressures of oxygen to carbon dioxide. At any given level of illumination, there will be a value of the ratio at which no net gas exchange takes place. This is the compensation point for carbon dioxide at that light fluence rate and partial pressure of oxygen. Below the compensation point (at lower pressures of carbon dioxide or at higher pressures of oxygen) the plant will have a net loss of carbon, and degenerate. This is the effective basis of the Warburg Effect: that oxygen inhibits photosynthesis in some plants. Those plants which show the Warburg effect have compensation points of the order of 40 parts per million of carbon dioxide in air, and belong to the C3 group, as opposed to the C4 group (Chapter 11) which have vanishingly low compensation points.

10.5 Relationship of Photorespiration to Other Pathways

It tends to be assumed that the oxygenase activity of RubisCO is not of any biological value, because:

(a) the photorespiration scheme of Fig. 10.2 has no positive effect, but only serves to diminish the loss of carbon from the pentose cycle, caused by the oxidase action of RuBisCO;

(b) RuBisCO preparations obtained from anaerobic photosynthetic bacteria show the same competition between oxygen and carbon dioxide, yet would never naturally encounter oxygen, implying that the oxygenase is an unavoidable concomitant of the carboxylase activity:

(c) the C4 and CAM metabolic processes (see Chapter 11) can be readily regarded as adaptations to mitigate the effects of the oxygenase reaction under environmentally stressful and therefore critical conditions.

Nevertheless it has been pointed out that the part of the process contained in the peroxysome is significant in the production of serine, from either glycollate or glycerate, and the chloroplast does indeed contain a phosphoglycerate phosphatase bound to the thylakoids (presumably light-regulated so as to avoid a futile cycle with glycerate kinase (195)), which could be of value in the production of serine.

A second positive value of the photorespiration system is protective. Isolated chloroplasts are rapidly damaged if they are illuminated in the absence of any electron acceptor. Probably molecular oxygen is reduced to a free radical such as superoxide which attacks the unsaturated fatty acids of the lipids. Intact plants, however, are able to survive periods when carbon dioxide fixation is severely restricted, for example when stomata are closed in bright light in order to avoid water loss. The photorespiration system then operates as a substrate cycle or 'futile cycle' ensuring that CO_2 is always available to consume NADPH and ATP, thus diminishing the likelihood of oxygen being reduced.

10.6 Experimental Avoidance of Photorespiration

It must be stressed that RuBisCO will always produce some phosphoglycollate if oxygen is present together with RuBP. In temperate (C3) plants up to half the carbon dioxide fixed may be lost in this way. Efforts have been made to find compounds that differentially inhibit the oxidase activity of RuBisCO with respect to the carboxylase; if this could be achieved in agriculture the rate of permanent carbon dioxide fixation could be increased by a factor of up to two. The economic effect would be a second agricultural revolution. No significant progress has been made to date, and it appears that the oxidase activity cannot be separated from the carboxylase. However, the output of carbon dioxide from the photorespiratory pathway can be affected by inhibitors. Thus α-hydroxysulphonates block glycollate oxidase, and allow the accumulation of glycollic acid. The compound glycidate (2,3-epoxypropionate) is an analogue of glycollate and glyoxylate, and inhibits the enzyme glutamate:glyoxylate aminotransferase in several leaves (197). Neither of these compounds offers much practical prospect of controlling the photorespiration pathway. Indeed the principle seems dubious; once the phosphoglycollate has been formed the need is to return the carbon (as much as is allowed) to the reductive pentose cycle as quickly as possible.

A search has also been made for mutants deficient in key enzymes of the photorespiratory pathway: the screening process involves finding seedlings that are sensitive to air, and need a high concentration of carbon dioxide for growth. Four enzymes of the photorespiratory system among others found to be deficient in mutants of barley were phosphoglycollate phosphatase, chloroplast glutamine synthetase, ferredoxin-dependent glutamate synthase (required for reincorporation of the ammonia released from glycine) and the enzyme responsible for the glycine–serine conversion. Although these lesions are in nuclear coded genes, the differential loss of the chloroplast glutamine synthetase with respect to the isoenzyme in the cytoplasm enabled the workers to show that the chloroplast ammonia fixation system is substantially independent of the cytoplasmic version (198).

Chapter 11

C4 Photosynthesis

11.1 The Warburg Effect—Exceptions

The observation, made by O. Warburg in 1920, that oxygen inhibited photosynthesis in *Chlorella*, applied to some other plants but not all. Certain tropical grasses were not affected. The susceptible group included the temperate plants, and the effect of oxygen can now be explained chiefly on the basis that oxygen competes with carbon dioxide in attacking the activated RuBP on the enzyme RuBisCO. The plants not subject to the Warburg effect, including the important 'grasses' sugarcane and maize, in some way avoid this competition between oxygen and carbon dioxide. As noted in Chapter 10, the oxidase and carboxylase activities of RuBisCO are virtually inseparable, and this result applies equally to the RuBisCO enzymes of both Warburg-susceptible and Warburg-resistant plants. Therefore we should look for an explanation in terms of a more successful competition in favour of carbon dioxide with respect to oxygen in the resistant group.

11.2 Compensation Points near Zero

An effect of the uptake of oxygen and output of carbon dioxide in the photorespiratory pathway is that at certain proportions of the two gases the net gas exchange by an illuminated leaf is zero, that is, the rate of photosynthesis is equal to the rate of respiration and photorespiration. Such conditions represent a *compensation point*. For temperate plants, at oxygen concentrations equivalent to air, the compensation point is found to be close to 40 parts carbon dioxide per million by volume. The compensation point does not change much over a wide range of light intensity, but increases with the oxygen concentration. In tropical grasses such as sugarcane and maize, compensation points have been measured in the range 0–10 parts carbon dioxide per million. It is suggested that even the low rates that are observed may not be normal respiration or photorespiration, but may rather be due to unusual processes such as the Mehler reaction, in which oxygen produced by PSII oxidises a low-potential product of PSI, generating hydrogen peroxide, and resulting in a net uptake of oxygen (see the review of Canvin (199)).

Attempts to demonstrate photorespiration in intact leaves of tropical grasses have failed. Not only is there no release of newly fixed carbon dioxide, which could have been an indication of an efficient recycling of carbon dioxide in the leaf, but there was a lack of turnover of the carbon in the glycine–serine pool. Also, in C3 plants the apparent quantum yield of photosynthesis could be increased by raising the concentration of carbon dioxide relative to oxygen, but in maize and sugarcane there was no effect. It seems likely that the compensation point of tropical grasses, for the purposes of comparison with that of temperate (C3) plants, is indeed close to zero. Tropical grasses have a means of suppressing photorespiration. It has to be borne in mind that maize and sugar-cane are particularly important annual crop plants, and have high rates of growth; sugarcane is regarded as the most efficient known example of biological solar energy conversion. It achieves under ideal conditions a dry-matter production with a heat of combustion representing more than 1% of the solar (PAR) energy input.

The key to the understanding of this efficient habit lies in a metabolic pathway known originally as Crassulacean acid metabolism, and recently styled C4 (for comparison with C3) photosynthesis.

11.3 C4-acid Formation in Crassulaceae: CAM

It has long been known that succulent plants, particularly the family Crassulaceae, accumulate acid in the cell vacuoles during the night, and lose it during the day. This phenomenon was detected by the sour morning-taste of the plants (see the review of Kluge (200)), and is due to malic acid, which may reach concentrations of the order of 100 mM in the vacuoles of the cells. Malate is actively transported across the tonoplast membrane into the vacuole, and the cytoplasm is effectively protected from the low pH.

During the day the malic acid disappears from the vacuoles, and the plant performs photosynthesis. These plants, of which the cactuses are spectacular examples, live in arid climates such as deserts, and are adapted for desiccating conditions. The leaves are often reduced in size, and the stomata are closed by day, preventing water loss but also preventing carbon dioxide entry. Photosynthesis takes place, producing oxygen, but with no apparent carbon dioxide uptake. At night, temperatures drop sharply, and the relative humidity therefore rises, making it safe for the stomata to open. One benefit of the water storage in these succulent tissues is the retention of sufficient heat for the nocturnal metabolic activity. This consists of the mobilisation and respiration of starch, and the fixation of carbon dioxide by the following four reactions:

$$\text{Pyruvate} + \text{ATP} + P_i \rightarrow \text{PEP} + \text{AMP} + PP_i \text{ (pyruvate phosphate dikinase)}$$

$$PP_i \rightarrow 2P_i \text{ (pyrophosphatase)}$$

$$\text{ATP} + \text{AMP} \rightarrow 2\text{ADP} \text{ (adenylate kinase)}$$

$$\text{PEP} + \text{HCO}_3^- \rightarrow \text{OAA} + P_i \text{ (PEP carboxylase)}$$

- -

$$\text{Net: Pyruvate} + 2\text{ATP} + \text{CO}_2 \rightarrow \text{OAA} + 2\text{ADP} + 2P_i$$

The formation of OAA above is effectively driven by two ATP molecules, and the uptake of carbon dioxide is further enhanced by the conversion of OAA to malate, which has an equilibrium greatly in favour of malate:

$$\text{OAA} + \text{NADH} \rightarrow \text{malate} + \text{NAD}^+ + \text{H}^+$$
$$\text{(malate : NAD dehydrogenase)}$$

This pathway appears to be located in the cytoplasm.

The diurnal breakdown of malate to provide carbon dioxide for photosynthesis may be achieved by two alternative means. One group of plants (the 'malic enzyme' group) oxidatively decarboxylate malate to pyruvate and carbon dioxide using either NAD^+ or NADP^+ as the coenzyme. The other group ('PEP carboxykinase') re-form OAA by means of malate dehydrogenase and decarboxylate OAA:

$$\text{malate} + \text{NAD(P)}^+ \rightarrow \text{pyruvate} + \text{CO}_2$$
$$+ \text{NAD(P)H} + \text{H}^+ \text{ (malic enzyme)}$$

$$\text{OAA} + \text{ATP} \rightarrow \text{PEP} + \text{CO}_2 + \text{NADH}$$
$$+ \text{H}^+ \text{ (PEP carboxykinase)}$$

It is likely (200) that the pyruvate is metabolised to starch by the gluconeogenesis pathway.

This cycle (Fig. 11.1) is known as crassulacean acid metabolism (CAM), and amounts to a separation in time between the nocturnal process of carbon dioxide capture by the plant, and the diurnal fixation by the reductive pentose cycle.

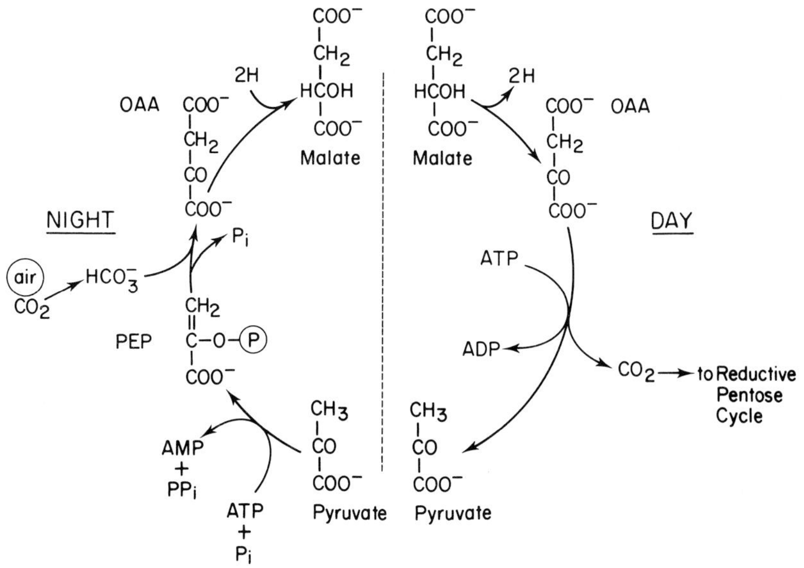

Figure 11.1. Crassulacean acid metabolism. The cells work a night-shift, concentrating CO_2 from the air into C4-acids. During the day the CO_2 is recovered. The separation in time makes sense of the otherwise futile cycle: pyruvate–PEP–C4–pyruvate

11.4 C4-acid Formation in Sugarcane

Kortschak et al (201) showed that application of the Calvin method of rapid $^{14}CO_2$ fixation followed by chromatographic separation of the products, in sugarcane, led not to PGA as the initial product but to the C4 acids OAA, malate and aspartate. Hatch and Slack found that the pathway of C4-acid formation made use of the same enzymes as CAM above (202), and subsequent researches led to the formulation of a pathway very similar to CAM, with the difference that aspartic acid formed by transamination from OAA was a prominent metabolite (Fig. 11.2).

PEP carboxylase uses hydrogen carbonate ions, in contrast to RuBP carboxylase which uses molecular carbon dioxide. The two enzymes therefore show different isotope effects (differences in the rate of reaction of isotopically labelled compounds related to the ratio of their masses): the naturally occurring $^{13}CO_2$ is discriminated against, with respect to $^{12}CO_2$, by 28.3‰ by the isolated RuBP carboxylase, and the carbon isotope ratio of whole C3 plants is consistent with that figure. C4 plants show a smaller discrimination, 13.56‰ in whole C4 plants, consistent with a discrimination of 9‰ by the isolated PEP carboxylase. The use of the mass spectrometer is of the greatest value in assigning plant species to C3/C4 categories (see the review by Troughton (203)).

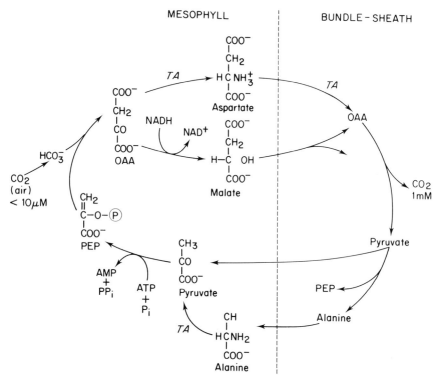

Figure 11.2. C4 photosynthesis. Instead of the temporal separation of Fig. 11.1, the cycle is divided between the bundle-sheath and mesophyll cells (see Fig. 11.4)

The pyruvate phosphate dikinase in C4 plants is located in the mesophyll chloroplasts, and operates via a form of the enzyme that is phosphorylated on a histidine residue. Burnell and Hatch (204) have shown that the activity of the enzyme is regulated by changes in light intensity, such that a 20-fold loss of activity occurs in the dark. This is achieved by phosphorylation on a threonine residue close to the histidine, which is carried out by a regulatory protein using the β-phosphate group of ATP, a very unusual feature. Even more unusual is the reactivation by phosphate, which removes the phosphate from the threonine, forming pyrophosphate, catalysed by the same regulatory enzyme (Fig. 11.3).

PEP carboxylase appears to have different isoenzymic forms, with different kinetic constants related to the habit of the particular plant. It might be expected that the low compensation point of C4 photosynthesis would be due to some property of the PEP carboxylase, the initial carbon dioxide fixing enzyme. Surprisingly, K_m values with respect to hydrogen carbonate or the associated concentration of carbon dioxide are if anything higher (the affinity is less) than

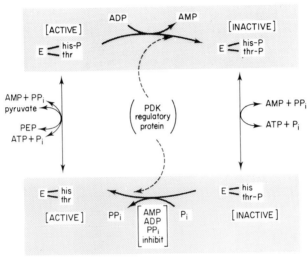

REACTION MECHANISM

$(\beta)(\gamma)$ (β) (γ)
E-his + AMP-P-P-P + Pi ⇌ E-his-P + AMP + PPi

(β) (β)
E-his-P + pyruvate ⇌ PEP + E-his

LIGHT–DARK REGULATION MECHANISM

Figure 11.3. Enzyme regulation in C4 photosynthesis. The example shown is the novel regulation mechanism proposed (204) for pyruvate phosphate dikinase (PDK). The action of PDK itself is shown in the two equations at the top, while the lower part of the Figure shows the action of the PDK regulatory protein, which catalyses both the activation and inhibition of PDK by reversible phosphorylation of a threonine group. It was suggested that the balance of regulation between the active (left side) and the inactive (right side) forms could depend on the concentration of ADP. Reproduced by permission of Professor M.D. Hatch and Elsevier Publications Cambridge. From Burnell, J.N. & Hatch, M.D. (1985) *Trends Biochem. Sci.* **10**: 289

the value of approximately 15 μM recorded for RuBisCO carboxylase, and it is not obvious how the low compensation point in C4 species is obtained.

11.5 Kranz Anatomy

Figure 11.4 shows a micrograph of crabgrass (*Digitaria* sp.), illustrating the features of Kranz anatomy, where the vascular bundles of the leaf are surrounded by closely touching bundle-sheath cells. Outside the single bundle-sheath ring (cylinder) there are loosely packed mesophyll cells with air spaces. The chloroplasts of the bundle sheath differ from the mesophyll: either they are considerably bigger, or they do not contain grana, depending on the species. Two photosynthetic cell types are always clearly present, and the process of C4 photosynthesis depends on their cooperation.

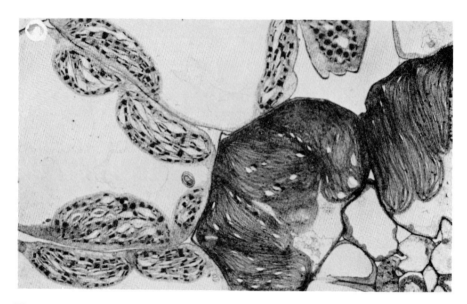

Figure 11.4.   Kranz anatomy, characteristic of C4 plants. Electron micrograph of a thin section of a portion of the leaf of crabgrass (*Digitaria sanguinalis* (L.) Scop.). The mesophyll cells (left) have large vacuoles and granate chloroplasts, and air spaces between them. The bundle-sheath cells surrounding the vascular bundle of conducting tissue (right) are almost filled by large chloroplasts with only rudimentary grana. Starch grains are present in both types of cell. Reproduced from Black & Mollenhauer (205) with permission

Mesophyll cells can be separated from the bundle-sheath strands by mechanical means, or by enzymic digestion of cell walls after stripping off the epidermis of the leaf. When the cell walls have been removed (two to three hours) the cell contents can be suspended in medium, yielding rounded protoplasts. The bundle-sheath strands are mechanically tougher, and digestion of the cell walls of maize strands separates the cells, without, however, destroying their rectangular appearance. The review of Edwards and Huber (206) describes the methods and means by which the integrity or viability of cells and protoplasts can be tested.

11.6 Principle of Carbon Dioxide Storage or Pumping

The C4 pathway may be written as a cycle (Figs 11.1, 11.2) which appears to show an overall carbon dioxide-dependent ATPase activity. The cycle is,

however, compartmented in the C4 tropical grasses between the mesophyll and
bundle-sheath cells, so that the C4 acid is formed in the cytoplasm of the
mesophyll, using chloroplast-derived ATP and (for malate) NAD(P)H. The
acid diffuses into the bundle sheath, where aspartate or malate are reconverted
to OAA which decarboxylates giving carbon dioxide and pyruvate. The pyruvate
(or alanine) diffuses back to the mesophyll. It is important to note that the first
photosynthetic fixation of carbon dioxide in the bundle sheath is carried out by
the normal reductive pentose cycle (i.e. C3). The bundle-sheath chloroplasts
(but not those of the mesophyll) contain RuBisCO and the other enzymes
of the pentose cycle. Isolated bundle strands or cells produce glycollate, and
photorespire in the same way as C3 plants. Production of glycollate in the intact
leaf, however, is reduced by more than 90%. The absence of photorespiration
in C4 plants can therefore be explained on the basis of, first, a very effective
competition by the increased level of carbon dioxide with respect to oxygen,
and second, recycling of any photorespiratory carbon dioxide, which is greatly
facilitated by the concentric structure of the Kranz anatomy.

In the succulents the C4 cycle is compartmented between night and day.
This is achieved partly by an internal clock, regulated by light, and partly by
the change in the distribution of metabolic effectors when the chloroplasts are
illuminated. The effect is that respiratory carbon dioxide as well as carbon
dioxide entering through the stomata of the leaf is trapped at night into C4
acids which are broken down on demand to supply the chloroplasts with carbon
dioxide. Whereas the C4 photosynthesis operates as a (spatial) carbon dioxide
pump, CAM is a (temporal) storage system. Once again, the final photosynthetic fixation of the carbon dioxide takes place by the normal reductive pentose
cycle. The C4 process is an addition to C3, not a replacement for it.

The control of the timing of the two phases of CAM is an example of a
circadian rhythm. Firstly, Queiroz (207) suggests that the connection between
the 'dawn signal' and the activation of the malic enzyme may lie in the clock-
controlled permeability of the tonoplast. Secondly, the seasonal dependence of
the rhythm may be related to the relative phase of the 'dusk signal' to the
circadian clock, so that the degree of activation of the PEP carboxylase step is
varied.

11.7 Variations

The appearance of C4 metabolism, whether as C4 photosynthesis or CAM,
seems to have occurred independently in several groups. There seem to be
three groups of C4 plants recognisable biochemically. The $NADP^+$–malate
group transport malate to the bundle sheath, where it is decarboxylated by the
$NADP^+$–linked malic enzyme; pyruvate returns to the mesophyll. Maize and
Digitaria are examples. The NAD^+–malate group transports aspartate, and
returns alanine or pyruvate to the mesophyll (example: *Atriplex spongiosa*).

Both malate groups transport (2H) as well as CO_2, and therefore diminish or eliminate the need for non-cyclic electron transport in the bundle sheath, which in turn diminishes the production of oxygen, further improving the preponderance of the carboxylase over the oxidase activity of RuBisCO. A partial or total deficiency of bundle-sheath PSII is often noted in these groups. The third group transport aspartate, which is decarboxylated by PEP carboxylase, and PEP is returned to the mesophyll (example: *Panicum maximum*). These three groups are not necessarily connected by evolution. In addition, it has been shown that in some plants there is a mixture of (direct) C3 and C4, and in others C4 photosynthesis or CAM behaviour can be induced by environmental circumstances. In some cases C4 photosynthesis has been observed in plants that do not possess Kranz anatomy, but in all cases there are two distinct green cell types present (discussed in the review of Ray and Black (208)).

An intriguing problem that may have a C4 explanation is the operation of the guard cell. Pairs of guard cells form the stomata in the epidermis of leaves; variations in light intensity and carbon dioxide levels cause changes in the shape of the guard cells and hence alter the stomatal aperture. In leaves with xeromorphic characters (thick cuticle, stomata restricted to the abaxial surface, etc.) there seems no doubt that closure of the stoma effectively stops water vapour loss and carbon dioxide entry. Partial closure may not have much effect, and in plants with thin cuticles gas exchange may not be particularly dependent on stomata anyway. Nevertheless stomatal opening and closing is still observed.

The guard cells contain chloroplasts whereas the other epidermal cells do not. However, the chloroplasts of the guard cells of C3 plants seem to be different from those of the adjacent palisade and spongy mesophyll cells. Biochemical differences between guard cell and mesophyll chloroplasts are a considerable lack of active RuBisCO in the former, together with the behaviour of the starch granules: the starch accumulates when the stomata are closed and is depleted during the time that they are open. Isolated guard cells are reported to have a rate of oxygen production 10–14 times that of carbon dioxide uptake, with the implication that the reducing equivalents produced by the light reactions are not used to operate the reductive pentose pathway. Isolated protoplasts of guard cells can be prepared by dissection of the epidermis and enzyme digestion of the cell walls, and it has been shown that the principal product of starch degradation is malate. This is consistent with the observed presence of PEP carboxylase as the main carbon dioxide-fixing enzyme. The guard cell appears to resemble the mesophyll cell of a C4 plant such as maize in that it produces malate in light. There is no conclusive evidence that the guard cell operates a substrate cycle with adjacent cells.

Guard cells have a blue-light effect, which actively pumps protons out of the cell. (This process may stimulate respiration, which is a known effect of blue light in algae.) Therefore an alternative view of the guard cell is that it regulates the stoma through a blue-light effect, photosynthesis being not directly involved.

11.8 Ecological Rationale and Possible Exploitation

The most apparent advantage of the C4 and CAM habits is that they conserve water. Water is lost through the same stomata by which carbon dioxide enters, and by only opening the stomata during the (cooler) night (CAM) or by greatly reducing the number of stomata and confining them to one leaf surface (C4) the loss of water is reduced, while the PEP carboxylase efficiently scavenges the carbon dioxide, increasing the diffusion gradient and hence compensating for the restriction on carbon dioxide entry. The loss of respiratory carbon dioxide is clearly avoided in CAM plants. C4 plants are able to counteract photorespiration in three ways. First, carbon dioxide produced during illumination is effectively retained within the bundle sheath; second, the raised concentration of carbon dioxide in the bundle-sheath cell diminishes the oxygenase activity of RuBisCO by competition; and third, in the case of the maize group, the diffusing C4 acid, being malate, carries reducing power as well as carbon dioxide and the bundle-sheath chloroplasts do not perform non-cyclic electron transport, and therefore do not produce oxygen; hence there is relatively little oxygen for the oxygenase.

The arid or semi-arid environments in which C4 plants flourish tend to have high diurnal temperatures. Photorespiration (in C3 plants) is greatly increased by elevated temperature.

C4 plants are spectacular performers in conditions of high temperature and high light fluence rate. Their rate of photosynthesis is saturated near full sunlight, at least an order of magnitude greater than the saturating intensity of C3 plants. Sugarcane is the most efficient converter of solar energy in biology. They are therefore most useful in agriculture in tropical and subtropical semi-arid regions. At the same time C3 plants are notoriously wasteful, and it has been estimated that cereal crops even in temperate climates, for example, may lose up to 50% of their fixed carbon dioxide by photorespiration. Discussion has therefore taken place over the advantage of breeding C4 characteristics into temperate plants.

Spartina townsendii has colonised European sand-dune habitats, and has C4 characteristics. However, it performs no better than adjacent C3 species. In general, given lower temperatures, and irregular insolation, C4 plants perform poorly; thus maize is a failure in Northern Europe. The critical enzyme responsible for the drastic inhibition by cold is the pyruvate phosphate dikinase.

It is also relevant that if the 50% loss in temperate plants could be avoided with the prospect of doubling the rate of effective photosynthesis, the leaf would be likely to outstrip the ability of the root to provide nutrients without corresponding enhancements. Existing C4 plants are more efficient in their use of nitrogen, and produce twice as much dry material per unit mass of nitrogen as C3 plants. The development of Kranz anatomy enables the C4 plant to maintain only half the quantity of RuBisCO per unit nitrogen, and the consequent economy allows C4 species to colonise nitrogen-poor soils.

Intermediate C3–C4 forms are known, for example *Panicum milioides*, which is intermediate in compensation point, anatomy and photorespiration. In air it behaves as a simple C3 plant, but the higher ratio of PEP carboxylase to Ru-BisCO suggests that near the compensation point it makes use of C4 characters.

Chapter 12

The Chloroplast Envelope and the Integrated Cell

12.1 Isolation of the Envelope

Figure 1.7 shows the envelope of the chloroplast as a double membrane. There is little apparent difference between the inner and outer envelope membranes in their appearance in the electron microscope, although they are very different in their biochemical and physical characters. They have different permeabilities, different complements of intrinsic proteins, different densities and different lipid compositions. Apart from their role in regulating the passage of solutes, metabolites and proteins, they have notably different roles in the synthesis of lipids, especially galactolipids (see Section 12.7.3).

Envelopes are isolated by careful rupturing of the chloroplast, either by osmotic shock (209) or by freeze-thawing (210), under careful conditions so as to avoid disruption of the thylakoid system. Membrane material from the envelope is separated from thylakoids and stroma proteins by means of centrifugation in a discontinuous density gradient, and the conditions may be chosen so as to achieve either sedimentation or flotation. Cline et al (210) resolved two membrane fractions in a linear sucrose density gradient. The densities were 1.08 and 1.13 g cm^{-3}, and the fractions were identified as outer and inner envelope membranes respectively, because proteins in the light zone were depleted when the chloroplasts were pre-treated with a non-penetrating proteinase, and the heavy zone contained the 29 kDa peptide identified with the phosphate translocator. The heavy zone contained relatively more of the galactolipid DGG (see

Section 12.7.3) and less of DDG with respect to the light zone, and the heavy zone also contained the enzyme galactosyl transferase.

12.2 Transporters

The envelope is a double membrane, with an E space between. The concept of transporter proteins, that is, proteins that combine with specific substrates and transport them across membranes, often in obligatory exchange for substrates going the opposite way, is somewhat difficult to visualise in this context. Transporters must either span the double membrane, which seems inherently unlikely, or be present in each membrane at the corresponding point so as to communicate directly, or the substrates must appear in solution in the E phase between two transportation steps, or one membrane must be leaky and therefore not require transporters for small molecules. The last is the situation in the mitochondrion, where the outer envelope is relatively permeable. Evidence that this is also the case in the chloroplast comes from the observation by Heldt and Sauer (211) that sucrose can cross the outer membrane and accumulate in the E space, but cannot enter the chloroplast. There is a group of proteins known as porins which make holes in membranes large enough for substrates up to 10 kDa in mass, and it is likely that they are responsible for the leakiness of the outer chloroplast envelope membrane. There is therefore no need for specific transporters for small molecules in the outer envelope.

Progress in this field has been more rapid since the development of the method of silicone oil centrifugation. A centrifuge tube is set up with a layer of perchloric acid solution on the bottom; the chloroplasts will eventually arrive in this layer and will instantly cease activity and be deproteinised. Above this layer is a layer of less dense silicone oil. In the simple form of experiment, a chloroplast suspension is placed over the oil, the radiolabelled metabolite to be studied is added at time zero, and at a predetermined time the tube is centrifuged so that the chloroplasts pass out of the solution, through the oil and into the acid. The chloroplasts carry virtually no aqueous medium through the oil layer. Under some conditions there may be a significant volume in the inter-envelope space (211) which can be controlled by parallel experiments with, say, sucrose. A refinement of the method (212) is to place the solution containing the metabolite of interest between two layers of silicone oil of different densities (adjusting the density of the middle aqueous layer as appropriate). The chloroplast suspension is placed above the top oil layer, and the centrifuge run up to speed. Depending on the speed, the chloroplasts pass through the middle aqueous layer in a precise time, during which they exchange solutes as their transporters permit. Most of the transporters act as antiports, and the entry of the target material will depend on the presence in the stroma of a compound that can leave on the same carrier. Therefore for some experiments the chloroplasts need to be pre-loaded with such a compound.

There appear to be two principal transporters in the envelope which are directly related to the reductive pentose cycle: one which exchanges any two of PGA, DHAP, GAP and phosphate, and a second which exchanges dicarboxylic acids such as malate, aspartate, glutamate, and 2-oxoglutarate. This set accounts for means by which ATP is exported from the chloroplast, since the actual rate of passage across the envelope by adenine nucleotides is slow, enough to provide for growth but not for the rates needed for active metabolic processes, which include the uptake of potassium ions or glucose by algal cells. In each case the uptake is light-dependent and inhibited by inhibitors of ATP formation. The intermediary formation and export of carbohydrate followed by respiration in the cytoplasm is ruled out by the continued light-dependent uptake of material when far-red light is used; the only effect of such light is to support cyclic electron transport (cyclic photophosphorylation) around PSI. ATP can be exported simply by the small loop shown in Fig. 12.1 involving PGA, BPGA, GAP and DHAP. The loss of (2H) units is avoided by the concomitant larger loop in which malate travels into the chloroplast stroma, where (2H) is released by the conversion of malate to OAA by means of malate dehydrogenase. Since OAA is present in very low equilibrium concentration with malate, it does not take part in translocation but is converted to aspartate. Aspartate is readily transported, but in order to avoid loss of nitrogen from the chloroplast a shuttle exists in which glutamate and α-ketoglutarate carry amino groups back into the chloroplast. (The principle of this complicated shuttle is also established for mitochondria in animal tissues.)

Evidence obtained from the silicon oil experiment suggests that there is an-

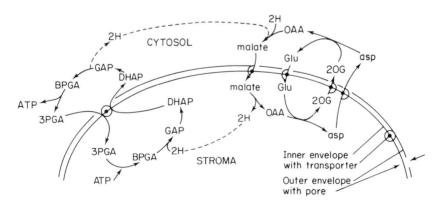

Figure 12.1. Shuttle system for the export of ATP from the chloroplast. The left-hand side shows the export of ATP by means of DHAP and the phosphate translocator; the right-hand side shows the recovery of (2H) and nitrogen, by means of the dicarboxylic acid translocator. The translocators are antiports, and exist in the inner-envelope membrane. The outer envelope contains pores formed by porins

other transporter for dicarboxylic acids excluding glutamate (212).

In all the cases above the transporters act as antiporters, that is, they only transport material A in one direction if they are able to transport material B (which may be the same as A) in the opposite direction. It is relatively easy to demonstrate B-dependent A-transport, and hence the existence of the antiporter. It is more difficult to show the presence of a uniport, and three approaches are (i) to show that the rate of transport saturates, that is, above a certain concentration difference the rate of transport is concentration-independent, or (ii) to show that there is competition between materials that use the same uniport, or (iii) to find an inhibitor of the transport process, such as phloridzin, the classical inhibitor of the glucose transporter in animal cells.

At first the passage of glycollate out of the chloroplast in the photorespiration cycle was considered to be a transport process, probably a uniport, but evidence was slow in coming and in 1980 it was a consensus that in fact both glycollate and glycerate crossed the envelope by diffusion. Recent work (196) has reversed matters again: it was shown that preloading the stroma with D-glycerate (or lactate) stimulated the uptake of glycollate. This phenomenon is known as trans-activation, and is found in glucose transport in the red blood cell. The rationale is that the carrier changes its conformation when it releases one metabolite and takes up the other, so that the presence of the second is necessary to minimise the turnaround-time. The protein reagent, N-ethylmaleimide (NEM), inhibits transport, as does L-mandelate (phenyglycollate). The inhibition by NEM is itself inhibited by glycollate, D-glycerate or L-mandelate, that is, the substrates protect the carrier against attack by NEM. Using radiolabelled NEM, Howitz and McCarty were able to identify a 35 kDa peptide in the inner envelope membrane as a possible glycollate–glycerate transporter (196). It was suggested that the carrier in one turn carries two molecules of glycollate with one proton in one direction (that is, outwards in photorespiration) and one molecule of glycerate in the reverse direction. Thus the ionic balance is preserved, and another example is provided whereby the two membrane crossing points of a transmembrane pathway depend on the same carrier (Fig. 12.2, compare Fig. 12.1).

The situation in CAM and C4 plants is not clear. These groups are not homogeneous. One problem is that the chloroplasts are able to take up pyruvate, and then export PEP by means of a PEP–phosphate antiporter. It may be supposed that the uptake of pyruvate is either a uniport, or, as in animal mitochondria, a diffusion process based on the appreciable lipid solubility of pyruvic acid.

Carbon dioxide passes across the envelope by diffusion. However, in *Chlamydomonas* (213) there is evidence of a light-dependent (uncoupler-sensitive) transporter of hydrogen carbonate in the chloroplast envelope, and this could be important at low levels of carbon dioxide. Cyanobacteria, which may be regarded as free-living chloroplasts, also possess such a system in their cell membrane. Algae in general and aquatic higher plants rely on hydrogen carbonate

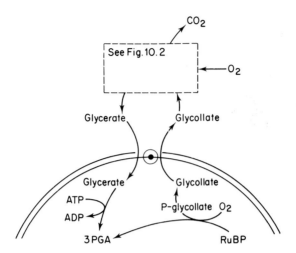

Figure 12.2.   Details of the glycollate–glycerate transporter (see Fig. 10.2)

accumulation, although in most cases studied the active transport is across the cell (plasmalemma) membrane, not the chloroplast envelope. The abundant enzyme carbonic anhydrase ($CO_2 + H_2 \rightarrow H_2CO_3$) has a role here in making the pool of hydrogen carbonate more rapidly available as carbon dioxide for RuBisCO. There is some evidence that in consequence some aquatic plants show markedly decreased rates of photorespiration; the increased CO_2/HCO_3^- supply (symbol C_i or C_{inorg}) effectively competes with oxygen at the RuBisCO site (see the discussion of recent papers, by G.J. Kelly (214)).

Chlorella was shown to store carbon dioxide in a form from which it was liberated after treatment of the cells with fluoride. This observation was at the time used by Warburg (215) to support his theory of the 'photolyte' (active carbon dioxide) split directly by chlorophyll, but it may in fact indicate a pool of carbamate stabilised by magnesium ions.

Large molecules also enter the chloroplast, of which the most studied, because of its great abundance, is the small subunit of RuBisCO. This peptide is synthesised in the cytosol, from mRNA of nuclear origin, on 80S ribosomes, and possesses, when first synthesised, a signal peptide at the N-terminus that adds some 3500 Da to its mass. This precursor is recognised by a specific receptor protein in the (inner) envelope as proposed (216) and later proved by R.J. Ellis (see reviews by Ellis (217, 218)). Such a mechanism is possibly a general mechanism for all the nuclear coded proteins of the chloroplast. In the case of the small RuBisCO subunit there may be an inner-membrane kinase that phosphorylates it, thus dragging it through by making the part in the stroma more hydrophilic. Recently, a specific receptor for the precursor of the small subunit has been claimed (219) to exist at points where the two envelope membranes are in contact. It may however be confused with the phosphate translocator (246).

12.3 Formation and Role of Sucrose

Virtually all vascular plants use the disaccharide sucrose to transport the carbon fixed by photosynthesis. Sucrose is osmotically active and the concentration in the phloem must be kept constant within limits, in spite of the obvious diurnal and random variation in the rate of carbon dioxide fixation, and the more steady but by no means constant rate of uptake of sucrose to form starch in storage organs. The synthesis of starch (see Chapter 9) met the immediate need of the chloroplast to regenerate inorganic phosphate and to limit the accumulation of monosaccharide phosphates that would otherwise result from carbon fixation by the reductive pentose cycle.

There is an elementary experiment known as the starch print. A plant is kept in darkness for two days, depleting its leaf starch, and one of its leaves is then exposed to sunlight for an hour through a photographic negative. The leaf is then boiled in 80% ethanol which removes the pigments and opens the starch grain structure, so that subsequent treatment with iodine produces a photographic positive print made up of the starch grains of the chloroplasts of the leaf. The high resolution of the print is evidence that starch is laid down where the carbon dioxide is first fixed, and that the rate of fixation even in the less illuminated parts considerably exceeds the rate at which carbon can be exported (as sucrose).

The chloroplast envelope possesses a vigorous transporter which exchanges triose phosphates (chiefly DHAP) for inorganic phosphate, in either direction. The triose phosphate and fructose 1,6-bisphosphate (F1,6BP) pools of the cytosol and stroma are therefore to be regarded as more or less continuous (the Gibbs Effect (see Fig. 9.3) indicates that equilibration is close if not exact). Control of the release of carbon from the combined pool is exerted at the point of fructose 1,6-bisphosphatase (the only other reaction that F1,6BP can undergo). This enzyme (F1,6BPase) is powerfully regulated by fructose 2,6-bisphosphate (F2,6BP), which reduces the affinity of the enzyme for its substrate (F1,6BP); see Fig. 12.3. Fructose 2,6-bisphosphate is formed from F6P by a specific kinase (F6P-2-kinase) and ATP, and is degraded by F2,6BPase to F6P again. The kinase is activated by F6P and by phosphate, and inhibited by PGA and DHAP; the phosphatase is inhibited by F6P and by phosphate. This is an example of a substrate cycle, in which an apparent F6P-dependent ATPase cycle provides, at a certain energy cost, a powerful regulation of the level of F2,6BP and hence of the enzyme activity of F1,6BPase. This results in a sigmoidal curve for the plot of F1,6BPase activity with respect to the concentration of the pool triose phosphate plus F1,6BP, and this means in turn that there will be very little drain of the pool (via F1,6BPase) until the pool concentration reaches a thresh-

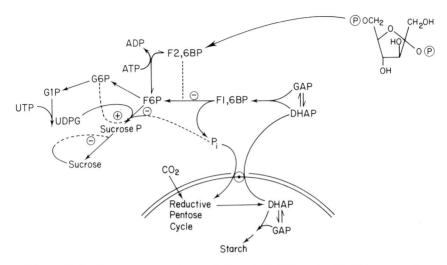

Figure 12.3.   Sucrose synthesis: regulation of F1,6BPase by F2,6BP (see text)

old, whereupon the enzyme activity will rapidly rise to a maximum. The time taken for these changes following the onset of illumination in the case of barley is of the order of 1 minute (220).

Fructose 6-phosphate is in equilibrium with G6P and G1P. Glucose 1-phosphate with UTP forms UDP-glucose and pyrophosphate; the immediate hydrolysis of the pyrophosphate to two phosphates makes the reaction irreversible. No regulation is known, but the concentration of UDP-glucose remains virtually constant irrespective of the rate of sucrose synthesis. The concentration of F6P is also kept constant, and therefore the combination of the two substrates to form sucrose phosphate by the enzyme sucrose phosphate synthetase can be presumed to be closely controlled.

Sucrose phosphate synthetase is activated by G6P and inhibited by phosphate. Other things being equal, an increase in the concentration of G6P and F6P will be accompanied by a decrease in phosphate. The result is that the activity of sucrose phosphate synthetase is sigmoidal with respect to its substrate F6P. The significance of the two sigmoidally controlled enzymes in the pathway is that as the supply of triose phosphate increases, first a supply of hexose monophosphate is made available to the general cell metabolism, and second, at greater triose phosphate levels, synthesis of sucrose phosphate begins.

Sucrose is released from sucrose phosphate by sucrose phosphatase, which is inhibited by its product, sucrose. The role of such inhibition in regulating glucose synthesis is discounted (221); pathways are usually controlled at their beginnings rather than at their ends.

In spite of its complexity, the above system of controls (Fig. 12.3) is insufficient to explain the diurnal behaviour of the starch–sucrose pattern of leaves of, for example, spinach. Sucrose formation (accumulation in the cell vacuole) begins before starch synthesis, and sucrose falls much faster than starch from their common maximum in mid-afternoon (at 16.00 hrs). The changes can be explained by an apparent increase in the activity of F6P kinase and a corresponding decline in F2,6BPase activity, irrespective of their allosteric effectors. By analogy with the same enzymes in liver, where protein phosphorylation under the influence of 5′ cyclic AMP produces the same changes, one might suspect protein modification in plants. Sucrose phosphate synthetase is also diurnally modulated, possibly by the same mechanism (220).

12.4 Carbon Partitioning between Starch and Sucrose

As seen in Chapter 9 the formation of starch is controlled at the enzyme ADP-glucose pyrophosphorylase:

$$\text{ATP} + \text{G1P} \rightarrow \text{ADP-glucose} + \text{PP}_i$$

by means of the activator PGA and the inhibitor phosphate. 3-Phosphoglyceric acid shares a translocator in the chloroplast envelope with phosphate and triose phosphates, but the pool of PGA is separated from triose phosphates by the reactions catalysed by PGK and glyceraldehyde 3-phosphate dehydrogenase. Cycling through these intermediates allows the export of ATP from the chloroplast. The stromal ratio of PGA/P_i is not easy to measure when the chloroplasts are inside the cells, but it is conceivable that it does not rise sufficiently for starch synthesis to begin inside the chloroplast until the formation of sucrose has reached saturation outside (in the cytosol).

12.5 The Integration of Carbon and Nitrogen Metabolism

Leaves contain nitrate and nitrite reductases; the latter is a chloroplast enzyme, and obtains its reducing power from ferredoxin. It is probable that nitrate reductase obtains reducing power from carbohydrate, as happens in roots. The supply of reduced nitrogen can be met by uptake of ammonia by the roots or the activity of nitrogen fixation in the nodules of leguminous plants. It has been pointed out (222) that nitrogen-fixing plants have a higher rate of carbon dioxide fixation, as if to feed the nodules, and tend to lose the leaf enzymes for reducing nitrate, with respect to control plants supplied with nitrate.

An ammonia cycle is also directly involved in the photorespiration cycle.

12.6 Chloroplast Repression of Mitochondrial Respiration

Much discussion has taken place concerning the constancy of mitochondrial respiration in the presence and absence of photosynthetic carbon fixation in the chloroplast. This point is essential for estimating the true rate of photosynthesis in dense algal suspensions. Probably the most reliable if indirect pointer is the Kok effect, which is observed in most algal and cyanobacterial suspensions, and is illustrated in Fig. 12.4. In effect Kok (223) measured the quantum requirement for carbon dioxide fixation at different light levels, including low levels where there was a net uptake of oxygen and output of carbon dioxide. (The intensity at which there is no net gas exchange is the compensation point for light, to be distinguished from the compensation point with respect to CO_2/O_2 balance.) Figure 12.4 shows a sharp discontinuity. Above the compensation point the value is constant and close to the expected eight quanta per molecule, while below it is constant and approximately four. At the time when the existence of the two photosystems was controversial, and O. Warburg was presenting evidence that the true quantum requirement was indeed four, the significance of the Kok effect was misunderstood. Later, the use of ^{18}O and ^{14}C (224) showed that as the light intensity increased from zero to the compensation point, photosynthetic and photorespiratory gas exchange increased together in proportion to the light, while respiration fell from its dark value to zero. Since, in algae, the rate of respiration is often about half the rate of photosynthetic gas exchange, the quantum requirement appeared to double.

The compensation point with respect to O_2/CO_2 has been measured as a function of the oxygen concentration. A straight line was obtained that could

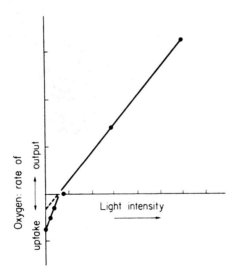

Figure 12.4. The Kok effect. The apparent increase in the quantum efficiency of photosynthesis, in *Chlorella* cells, on increasing the light intensity from below to above the compensation point. The effect is due to the suppression of mitochondrial respiration. Reproduced by permission of Elsevier Science Publishers. From Kok, B. (1949) *Biochim. Biophys. Acta* **3**: 626

be extrapolated through the origin. This is only possible if gas exchange at or above the compensation point was entirely due to the chloroplast; the continued activity of mitochondria would have produced a curve. Therefore, mitochondria are probably inactive during photosynthesis.

There is no reason to suspect a direct connection between the respiratory and photosynthetic electron transport chains of eukaryotes. However, in prokaryotes such as cyanobacteria and purple non-sulphur bacteria there is a cytochrome c-oxidase that is active in light and darkness; in the former group the rate is too low to make much difference, and in the latter photosynthesis is suppressed by oxygen, and indeed one function of the oxidase may be to act as an oxygen 'getter', removing traces from the medium.

The mechanism by which the chloroplast regulates the mitochondrion has been suggested to be the $NAD^+/NADH$ ratio in the cytoplasm, as well as the ATP/ADP ratio and the supply of phosphate. In effect the respiratory system is thrown into the 'resting' state. It has been shown that the labelling of citrate-cycle acids is very slow while chloroplasts are fixing radiolabelled carbon dioxide, suggesting that glycolysis is inhibited.

12.7 Development of Chloroplasts

12.7.1 Origin of the chloroplast

The origin of chloroplasts in the development of a plant is a hard problem, principally because of the impossibility of observing fertilised ovules or meristematic cells for long enough to recognise which of the many granules develop chlorophyll. In *Nicotiana* and in spinach, Wildman and Jope (225) observed that chloroplasts are absent from both ovules and pollen grains. The developing shoot is an apical meristem which begins to form leaf primordia. The meristem cell does not contain any particle in which chlorophyll can be detected by the sensitive fluorescence method, but such particles appear at an early stage in the proliferation of cells by mitosis. The chlorophyll-containing particles are about the size of one granum (0.5 μm), and increase in size as the cell undergoes mitotic divisions. New grana-sized chloroplasts appear during this stage.

The above authors pointed out that the chloroplast is entirely maternally inherited. Chloroplast DNA must be present in the ovule. In variegated leaves, there are two populations of chloroplasts, one of which fails to become green. Since variegation is transmitted maternally, there must be at least two copies of the chloroplast genome, one normal, the other deficient, in the ovule. There is a relationship between the volume of a chloroplast and its DNA content; extrapolation back to the 0.5 μm primordium indicates one DNA per chloroplast at its inception. This in turn prompts the speculation that each new granum in the growing chloroplast results from each new DNA copy.

So far as can be seen, the earliest primordia are double-membrane vesicles that have been claimed to originate as outfoldings of the nuclear outer membrane, on the basis of observations where amoeboid plastid-like organelles were seen to fuse reversibly with the nuclear envelope. The distinctive composition of the chloroplast envelope makes the suggestion difficult to accept at present. The origin of the thylakoid system within the developing plastid is similarly difficult. The claim that presumptive thylakoid membranes grow into the vesicle from the inner envelope depended on micrographs in which there were elongated intrusions from the inner envelope into the stroma. Again, given the distinctive composition of the membranes, the hypothesis must be treated with reserve.

Since chloroplast DNA differs from the nuclear type, resembling that of prokaryotes, it is unlikely to be assimilated into the nuclear (eukaryotic) genome during its cryptic phase. One is therefore looking for an event in the sequential development control part of the nuclear genome that activates the expression of the chloroplast genes. This is of great current interest as a general problem in molecular biology.

Many higher plants do not develop chlorophyll in the dark, and the plastids form etioplasts, in which the characteristic structure is the prolamellar body. This is a three-dimensional lattice of tubular membranes, in which the main protein is protochlorophyll holochrome. A very brief exposure to light is sufficient to cause the initiation of the conversion of protochlorophyllide to chlorophyll, and synthesis of the thylakoid proteins. Probably the increasing quantity of protein in the membrane is sufficient to reduce the curvature and bring about the normal appearance of thylakoids.

Gymnosperms kept in the dark commonly form chloroplasts containing functional PSI and PSII complexes, that is, with a normal pattern of chlorophyll–protein complexes, but they lack an active oxygen-evolving system until stimulated by light. Generalising, genes coding for proteins which need light to appear in the chloroplast are termed photogenes. The maize photogenes are grouped together in four small regions of the circular DNA, and most are related to polycistronic RNA transcripts which increase in proportion to the total RNA transcript when dark-grown seedlings are illuminated. Research has been directed at the particular promoter which may be directing the light effect.

The pigment most likely to be the photoregulator is phytochrome, at least when there is virtually no chlorophyll. Phytochrome activates nuclear genes for chloroplast proteins as well. Chlorophyll may also be regarded as a photoactivating pigment, acting indirectly through a photosynthetic product. Control is exerted at both the transcription and translation stages.

12.7.2 Synthesis of chlorophyll

Shemin, in 1940, showed that the tetrapyrrole ring system was formed from δ-aminolevulinic acid (δALA) in the blood of ducks (where the red cells continue to synthesise haemoglobin). This was one of the first uses of the isotope ^{14}C

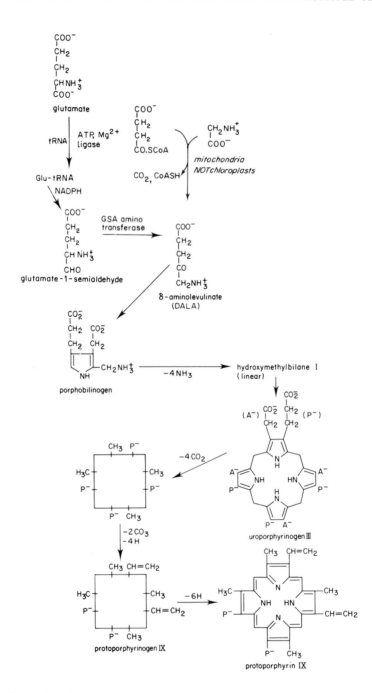

Figure 12.5.   (*see legend on opposite page*)

to elucidate a metabolic pathway. δALA was formed from succinyl CoA and glycine, and the Shemin pathway was later established as the pattern for all haem groups. The synthesis of δALA in the chloroplast stroma, however, differs in that all the five carbon atoms of δALA originate from glutamate (226).

The glutamate was shown by Schön et al (226) to be attached to a specific transfer RNA molecule, but instead of proceeding to the ribosome in the expected way the glutamate is reduced by means of $NADP^+$ to glutamate-1-semialdehyde. A transamination reaction forms δALA.

The common pathway ends with the formation of protoporphyrin IX (Fig. 12.5a), and instead of incorporating iron to continue towards haem, Mg^{2+} is complexed, and the pathway proceeds as shown in Fig. 12.5b. Protochlorophyllide accumulates in etioplasts, and light is required (absorbed by the protochlorophyllide) for the reduction using NADPH that forms chlorophyllide.

Chlorophyllide needs esterification with the alcohol phytol at position 17^3 to become Chl a. A long-standing problem has been cleared up: the only enzyme previously known to operate on this ester site was chlorophyllase, which is only active in aqueous organic solvents, and removes phytol from chlorophyll, forming chlorophyllide if the solvent is acetone, or (e.g.) ethyl chlorophyllide (alkyl exchange) in ethanol. Consistent with the rule that hydrolytic enzymes are catabolic, and that condensation reactions are carried out with activated precursors, it has been shown that the initial product is the geranylgeranyl ester, from the precursor GG-pyrophosphate; the pyrophosphate is a leaving

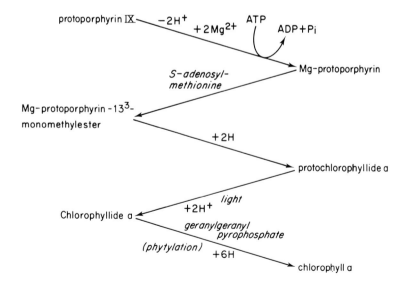

Figure 12.5. Biosynthesis of chlorophyll. Unlike animal cells, chloroplasts do not incorporate carbon from glycine into the tetrapyrrole ring

group in the esterification (227). Reduction of three (out of the four) double bonds converts the geranylgeranyl group to phytyl in situ. The geranylgeranyl group is an intermediate in the biosynthesis of carotenoids shown in Fig. 12.6.

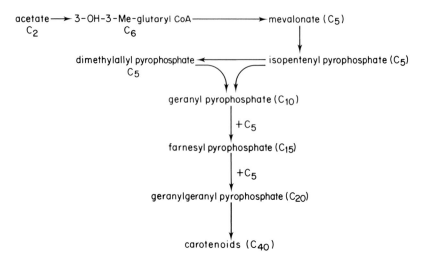

Figure 12.6.   Biosynthesis of carotenoids from acetate

12.7.3 Synthesis of chloroplast lipids

The extensive membrane system of the photosynthetic thylakoids contains polar lipids amounting to some 40% of its dry weight. About half of this is accounted for by diacylgalactosylglycerol (DGG)(also known as monogalactosyldiacylglycerol, MGDG); the formulae of lipids discussed in this section appear in Fig. 12.7. There is a second galactolipid present, diacyldigalactosylglycerol (DDG)(digalactosyldiacylglycerol, DGDG), and these together make up 70% of the thylakoid lipid. The remaining lipids of the thylakoid are phosphatidylglycerol (PG), diacylsulphoquinovosylglycerol (DSQ), phosphatidylcholine (PC) and other phospholipids.

DGG is the subject of a useful review by Gounaris and Barber (228) and is described there as the most abundant polar lipid in Nature. It is not a single compound, however, but a group, the members of which differ in respect of the nature (C_{16} or C_{18}) of the fatty acids, their position on the glycerol, and their degree of unsaturation. The same is true of all the groups described above, but the variation is seen to be restricted when it is set out in Table 12.1 It may be noted that unsaturated galactolipids appear in the chloroplast envelope also, although the proportions of DDG and DGG are reversed. Elsewhere in the cell, e.g. mitochondria and endoplasmic reticulum (microsome fraction),

	Structure	Formal charge at neutral pH	% of total acyl lipid
Monogalactosyldiacylglycerol	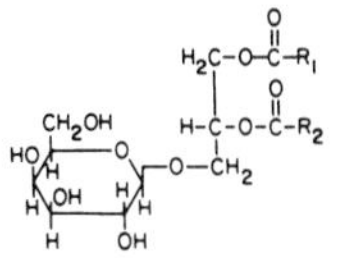	neutral	40
Digalactosyldiacyglycerol	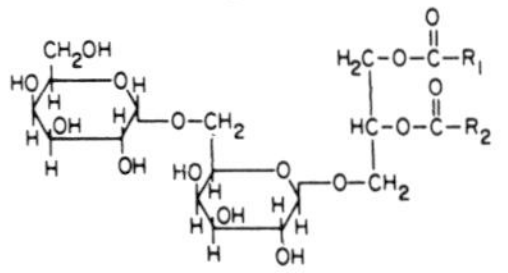	neutral	30
Diacylsulfoquinovosylglycerol	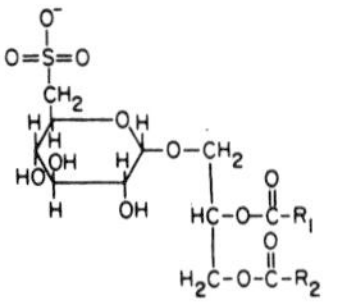	anionic	10
Phosphatidylglycerol	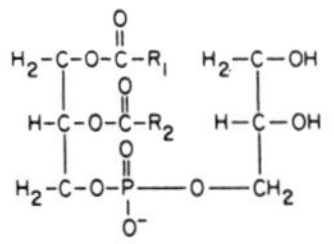	anionic	15
Phosphatidylcholine	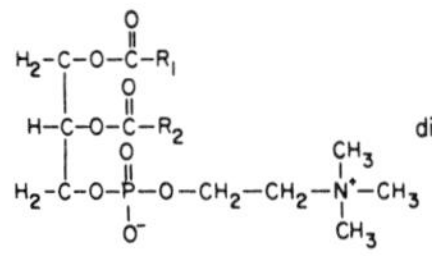	dipolar ionic	5

(a)

Acyl groups:

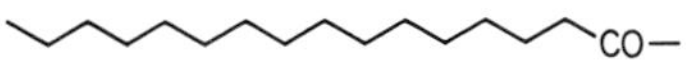 CO—	C_{16} 16:0	palmitoyl
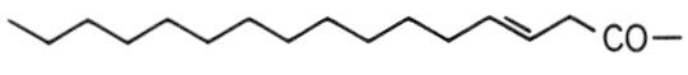 CO—	C_{16} 16:1 t	Δ_3-trans-hexadecanoyl
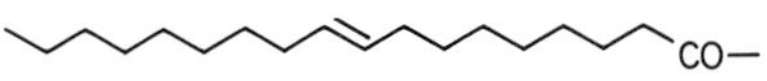 CO—	C_{18} 18:1	oleyl
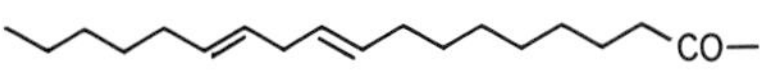 CO—	C_{18} 18:2	linoleyl
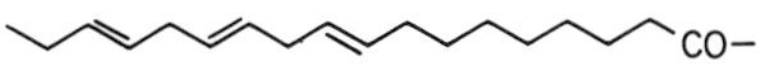 CO—	C_{18} 18:3	linolenyl

(b)

Figure 12.7. The major lipids of the thylakoid. (a) Formulae of five types of lipid. (b) Formulae of five types of fatty acid, corresponding to the R-groups in (a)

Table 12.1 Acyl lipid and fatty acid molecular species composition of typical higher plant photosynthetic membranes

Lipid molecular species	DGG	DDG	DSQ	PG	PC
18:3/18:3	92	70			
18:3/16:0	4	24	62	2	51
18:2/16:0			33	5	34
18:1/16:0				7	8
any $C_{18}/16:1^3t$				45	
16:0/16:1^{3}t				37	
Total	38	29	12	14	6

18:3, linolenoyl; 18:2, linoleoyl; 18:1, oleoyl; 16:0, palmitoyl; 16:1^{3}t, Δ_3-trans-hexadecanoyl. These values are averages of data from a range of higher plant species, including maize, spinach, oleander, and soybean. From Murphy, D.J. (1986) In *Photosynthesis III. Encyclopedia of Plant Physiology*, new series, vol. 19, Springer-Verlag, Berlin, p.714, with permission

the principal lipids are phospholipids with mainly saturated fatty acids, and in that respect they resemble their animal counterparts.

In a stimulating review, Roughan and Slack account for some of this variation by considering the synthesis of lipids in the cell (229); the chloroplast is the single source of fatty acids, and the galactolipids are elaborated in the inner membrane of the chloroplast envelope. This is in contrast to animal (e.g. liver) cells where fatty acid synthesis is a cytosolic function. The fatty acids are built up by addition of C2 units (from malonyl CoA), to the growing fatty acid chain carried on the acyl-carrier protein (ACP) as far as the stage of palmitoyl-ACP. The notation used is (16:0) for palmitic acid, meaning a 16-carbon acid with no double bonds. 16:0 is extended to 18:0 (stearoyl ACP) by the enzyme palmitoyl ACP elongase, and 18:1 (oleoyl-ACP) is then formed by stearoyl-ACP desaturase. Two acyl-transferase enzymes transfer fatty acids from the ACP to glycerol-3-phosphate. Frentzen et al showed that one was soluble and was specific for the C1 of glycerol phosphate, while the other was located in the inner envelope and was specific for the 2-position (230). The soluble enzyme had a preference for 18:1, the second for 16:0. The authors concluded that this system would tend to produce mainly 18:1/16:0 phosphatidic acid (PA), with some 16:0/16:0; the notation indicates the fatty acid type esterified on positions 1 and 2 of the glycerol. The pattern 18/16 is found in cyanobacteria, and lipids containing glycerol esterified in this way (18/16 and 16/16) are known as prokaryotic lipids. 18/18 and 16/18 are termed eukaryotic. Lipids of higher-plant chloroplasts can contain both prokaryotic and eukaryotic diacylglycerol, and a classification exists for higher plants based on the relative occurrence of 16:3 and 18:3 acyl groups in the various classes of lipid. Chloroplast PG (see

Table 12.1) always preserves its prokaryotic character.

The inner envelope is able to hydrolyse PA to diacylglycerol (DG), and to transfer galactose from UDP-galactose to the vacant 3-position of the glycerol. The UDP-galactose presumably comes from the cytosol. The result is DGG. The fatty acyl residues are unsaturated at positions 9, 12 and 15 to give 18:3 and 16:3; the geometry of the double bonds is cis, although there is a minority component of the 16:1 form that has a trans unsaturation at position 3. The 18:1–3 forms are oleate, olenate and linolenate respectively and are important in mammalian nutrition. The mechanism of unsaturation is not known.

The above scheme applies to spinach, which is classified as a 16:3 plant, i.e. it has predominantly prokaryotic lipids. It does not seem to work in isolated intact chloroplasts from peas, however, which are eukaryotic, and it appears that the synthesis stops at DG (prokaryotic). DG does not become incorporated into DGG in the inner envelope. DGG is formed in the outer envelope of pea chloroplasts, but not from prokaryotic DG (231). Roughan and Slack showed that eukaryotic DG was synthesised on the endoplasmic reticulum. Their hypothesis suggested that fatty acid leaves the chloroplast stroma as acyl CoA, passing through both envelopes into the cytosol. They are incorporated into predominantly 18:1/18:1 DG that are unsaturated further to 18:2/18:2 before being transferred to the outer envelope, where they are galactosylated and unsaturated to the final 18:3/18:3 state, again by poorly understood mechanisms. The outer envelope has an interlipid galactosyl transferase that produces DDG from DGG. It is a difficult question how the lipids are moved from membrane to membrane.

The cyanobacterium *Anabaena variabilis* incorporates glucose into DG, which is epimerised to a galactosyl residue before the acyl chains are unsaturated.

PG in the chloroplasts of higher plants of all groups is of the prokaryote type, and the 2-position carries 16:1 (trans-)3-hexadecanoic acid. This accounts for the DG produced in the chloroplast envelope in peas. Envelope membranes form CDP-DG from DG and CTP, and hence PG from CDP-DG and glycerol phosphate.

The above discussion and Table 12.1 may give the impression that plants are divided between two groups represented by peas and spinach. There is in fact a more continuous distribution when DGG, DDG and DSQ are considered, since in the 16:3 group of plants the DG from both sources mixes in the envelope membranes. The proportions are species-specific.

Functions of chloroplast lipids

There is a consensus that the thylakoid membrane is a bilayer. This is not easy to see on electron micrographs of thin sections after fixation and staining, but it is necessary in order to understand the freeze-etch micrographs in which the two leaflets of the thylakoid are split apart. Bilayer unit membranes are essentially lipid (polar lipid, usually phospholipid) in nature, and the first function of lipids in the thylakoid is to maintain this structure. DDG and DGG are paradoxical, since DGG has been shown to form a liquid-crystalline hydrated phase in which the DDG molecules pack into cylinders, with the polar heads towards the axis in each cylinder and the cylinders arranged parallel in hexagonal packing. DDG is therefore described as a non-bilayer lipid. Its behaviour depends on the acyl chains being unsaturated, and if some of the double bonds are artificially hydrogenated the modified DDG forms lamellae.

The thylakoid membrane has a relatively high content of protein, and one suggestion is that DDG by virtue of its conical structure is able to pack an imagined spherical protein complex more neatly into an existing bilayer. The cartoon in Fig. 12.8 illustrates this concept. In this sense the protein stabilises the bilayer of the membrane, rather than the other way round. There may be a useful analogy with the mitochondrial cristal membrane, which also has a high protein content and a non-bilayer lipid, cardiolipin.

It is also a consensus that the thylakoid membrane is very fluid. Extracted lipids form structures in which fluidity at the microscopic scale can be examined by means of fluorescent probes. Fluorescence retains its polarisation so long as (i) no exciton transfer takes place and (ii) the period of rotation of the fluorescent molecule is long in relation to its fluorescent lifetime. By this means it has been shown that extracted lipids are very fluid, and even the natural protein-containing thylakoid is relatively fluid compared with other biological membranes. Fluidity is needed for the shuttling of PQ from PSII to the *bf* complex; this distance may be considerable given the lateral separation of the two. It is also required for the movement and reorganisation of the LHCII and possibly other proteins following their reversible phosphorylation, as in the state I to state II transition. Thirdly, when grana are unstacked the protein complexes are seen (1) to move readily round the membranes, and to resegregate themselves when stacking is reimposed.

Fluidity is stated to depend on lipid composition, unsaturation of acyl chains and temperature. However, alteration of lipid composition and unsaturation by addition of extraneous lipids and hydrogenation has relatively little effect (see the review of Siegenthaler and Rawyler (232)). The chief determinants of fluidity appear to be the temperature and the lipid–protein ratio, since it is observed that plants adapted for chilling (228) have a higher ratio of lipid. It would appear that DDG is an inherently mobilising lipid. Why is there a different pattern of DGG–DDG in the envelope? The envelope may not need to be any more fluid than, say, the endoplasmic reticulum, if its main function is to be a galactolipid factory.

Considerable debate is in progress over the question of the need for localised regions of membrane containing different lipid complements for the different functions of the thylakoid. The obvious starting point is the separation of stroma

Figure 12.8. An impression of the specific role of galactolipids in the construction of the thylakoid membrane. Reproduced by permission of TAB and Elsevier Publications Cambridge. From Gounaris, K. & Barber, J. (1983) *Trends Biochem. Sci.* **8**: 378

thylakoids and grana vesicles, which can be achieved by means of mechanical shear followed by partition between aqueous immiscible phases containing polyethylene glycol and glucan, and avoiding any use of detergents. Differences are found, but the results can be normalised with respect to total lipid, to protein or to chlorophyll, and it might appear that the chief difference was a consequence of the agreed separation of PSI in the stroma thylakoids from PSII and LHCII in the grana. The conclusions with respect to lipids are that DSQ and PC are more abundant in stroma thylakoids, and DDG in grana (232).

The lipid distribution has also been examined with respect to the inner and outer leaflets and surfaces of the thylakoid. Antibodies against specific lipids have revealed more antigen at the inner surface, DSQ and phospholipids at the outer surface, and galactolipids in the inner leaflet. (Most galactolipids did not react and were therefore presumed to be internal.) The antigenic fraction of DDG was closely associated with the coupling factor (F_0). These studies were aimed at a possible functional role for certain lipids, that is, a suggestion that particular protein complexes might require for their activity a particular lipid environment.

Isolation of complexes (LHCII, PSII, PSI, b_6f, CF) involves the use of detergents, and the more purified the product, the more likely it is that the detergent has replaced the intimate lipid. Nevertheless there is a wide choice of detergents and some lipid is often found in the preparation. Thus LHCP (prepared by means of anionic detergents) has more PG and DSQ and less DDG than the average membrane, while PSI (CPI + anionic detergent) contained DSQ and phosphatidylinositol (PI). With different detergents and preparative methods other results are obtained, but nevertheless the association of DSQ with PSI and PG with LHCII (at least in a monomeric form) seems to be often found.

The antibodies against the lipid classes mentioned above were tested for their inhibitory effect on PSI and PSII photoreactions: with decreasing effect, PSI was inhibited by anti-PC, DDG and DGG, and PSII by anti-DGG, PG, DSQ, PI and PC (232, 233). Clearly there are good grounds for expecting specific associations between groups of lipids and protein complexes, at least when the lipids are performing a structural role. The fact that complexes can be prepared in which the lipid composition varies while the activity remains the same tends to suggest that the role of the lipids is only structural, and there may be no functional role played by lipids.

Further Reading

The metabolic interrelationships of chloroplast and cytoplasm are emphasised in the following text:

Halliwell, B. (1984) *Chloroplast Metabolism: The Structure and Function of Chloroplasts in Green Leaf Cells*, revised edn, Clarendon Press, Oxford.

Chapter 13

The Impact of Molecular
Genetics on Photosynthesis

The workings of photosynthesis depend on proteins, which act variously as supports for the pigments, electron-transfer carriers, transporters and enzymes. The subject of molecular genetics is the relationship between the proteins occurring in a given situation and the sequences of DNA, the genes, which govern their structure and production. The genetic approach has brought insights into a wide variety of biological problems, and will undoubtedly increase in importance. With regard to photosynthesis, it is proving easier to determine the sequence of amino acids in a protein indirectly, from the sequence of the gene, than directly by the classical biochemical methods. Secondly, the chloroplast genes in both nucleus and chloroplast coordinate to produce components of multi-subunit proteins, and in all photosynthetic systems the production of the photosynthetic apparatus is under the control of factors which operate at the genetic level. Thirdly (this is not an exhaustive list) plants are adapted to widely differing environments, by means of greater or lesser variations in their structures: genetic analysis has provided insights as to how this variation has arisen.

The object of this chapter is to deal as briefly as possible with the inevitable technical terminology of the subject, to summarise the methods employed and some of the findings and, rashly, to discuss the scope and limitations of the genetic approach.

13.1 Genomes

A genome is a molecule of DNA containing genes. Eukaryote cells have nuclei, which contain chromatin (a DNA–protein complex); at the time of cell division, the chromatin condenses and can be seen, after staining, in the light microscope as a set of chromosomes; the set is known as the nuclear genome, as opposed to other DNA, the genomes of organelles such as chloroplasts and mitochondria. In bacteria DNA does not occur as chromatin and cannot be seen as visible chromosomes. Nevertheless the major DNA molecule, always present in the cell, is termed the chromosome, as opposed to other (smaller) DNA molecules such as plasmids which may come and go.

The concept of the gene has undergone considerable change since the time of Mendel. The sheer size of the genomes of cells and the complexity of the information contained has generated a specialist vocabulary. In simple terms, a gene is a section of DNA that recognisably codes for a specific molecule of ribosomal RNA (rRNA), or transfer RNA (tRNA), or messenger RNA (mRNA). The simple usage becomes more complicated in view of the fact that the coding regions have associated controlling sequences, and also that sequences identifiable with the amino acid sequence of a known protein may only represent part of the DNA sequence for a mRNA molecule. An alternative concept of a gene is an identified sequence of DNA in which changes (mutations) are expressed as a change in a known cell constitutent. Clearly there is a subjective element in the definition of any particular gene which depends on the present state of knowledge .

In virtually all genomes there is a considerable proportion of DNA which is not known to code for anything; this proportion is greater in nuclear genomes, and increases (with some exceptions) with the ascent of the presumed evolutionary tree.

The bacterial (prokaryote) genome forms a continuous loop, and is often described as circular, but of course collapsed in shape, as can be seen in electron micrographs when a bacterium is carefully disrupted. The contour length can be measured on the micrographs.

Chloroplast DNA is of the prokaryote type, in that it is a continuous loop and does not form a chromatin complex. It is, however, smaller than the genomes of most bacteria. The circles of DNA from higher plants are 44–46 μm in total contour length (much longer than the longest dimension of the chloroplast). Algal chloroplast DNA is more variable in size. A length of 45 μm corresponds to about 150 000 base pairs, or a molecular mass of 100 MDa. There are several hundred copies of the DNA circle in a mature chloroplast, so that the total DNA content is greater than that of an average bacterium; the many chloroplasts in a leaf cell together contain DNA comparable to the quantity in the cell nucleus.

13.2 Transcription

DNA in the genomes of chloroplasts, nuclei and bacteria is double-stranded. Each strand has a recognisable direction, as defined by the phosphodiester bond from the 5' position of the deoxyribose of one nucleotide to the 3' position of the next. The two strands are hydrogen-bonded so that they run in antiparallel directions. Because of the base pairing, in which the base adenine (A) pairs with thymine (T), and guanine (G) with cytosine (C), the sequence of one strand governs the sequence of the other, which is said to be its complement (Fig. 13.1).

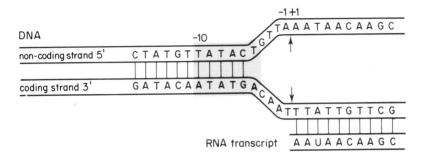

Figure 13.1. Relationship between the two strands of duplex DNA and mRNA. The coding strand of the duplex DNA (lower) separates from the complementary (base-paired) non-coding strand (upper) and the mRNA is synthesised as the complement to the coding strand. Thus the mRNA corresponds with the non-coding strand of the DNA (except for the substitution of U for T). This example shows the DNA sequence from the promoter region of the psbA gene in wheat chloroplasts (234), and the conserved −10 region is shaded. The figure +1 on the non-coding strand indicates the beginning of the transcribed sequence

DNA in whatever location is transcribed by the action of RNA polymerase in concert with enzymes that open the double helix. RNA polymerase recognises and attaches to sections of DNA known as promoter sequences, and begins transcribing, that is, synthesising RNA, by assembly of complementary nucleotides starting at the adjacent initiation site. It is an important detail that transcription necessarily proceeds from the 3' to the 5' direction of the DNA strand being transcribed, and the RNA is assembled, growing from the 5' to the 3' direction. The RNA transcript is therefore identical to the non-transcribed strand of the DNA (except for the ribose and the use of the base uracil instead of thymine). It is convenient to discuss the structure and sequence of genes with reference to the RNA transcript, and therefore to the non-coding strand of DNA. Thus the promoter sequence is on the 3' side of the section of DNA to be transcribed, but by convention it is described in terms of its complement, which is on the 5' side of the complement to the section to be transcribed (Fig. 13.1).

The promoter sequences of prokaryotes such as *E. coli* are very similar to those of chloroplast DNA, and the RNA polymerase isolated from the bacterium is able to recognise chloroplast promoters. These sequences are contained in the region up to 40 bases on the 5′ side of the transcription initiation site and are particularly critical around the −10 and −35 positions. There are some differences, however (234).

Beginning at the initiation site, the RNA is transcribed so that its sequence is governed by that of the DNA. Transcription continues until a sequence recognised as a termination sequence is encountered. The RNA at this stage is termed a primary transcript. The primary transcript is modified by a certain amount of clipping at each end, producing the definitive rRNA, tRNA and mRNA molecules. In bacteria and chloroplasts, one primary transcript may be cut up to yield several different DNAs. In addition, the primary transcript contains 'intervening sequences' or 'introns', rarely in bacteria and chloroplasts, but very commonly in nuclear genes. Introns do not, in general, code for any useful RNA, and are cut out of the primary transcript. The sequences that are expressed ('exons') are spliced together, and mRNA from the nuclear genome, but not that from prokaryotes or chloroplasts, is further modified by the addition, to its 3′ end, of a 'tail' of up to 200 polyadenine units.

13.3 Translation

The process of synthesising protein, known as translation, involves the ribosome. Ribosomes are particles containing approximately equal parts of rRNA and protein. Bacterial ribosomes are very similar to those of chloroplasts, and have sedimentation coefficients of 70S, whereas cytoplasmic ribosomes of eukaryotes sediment at 82S. The 70S ribosomes are composed of a large (50S) and a small (30S) subunit, and the predominant rRNA in each is 23S and 16S respectively. There are two additional rRNA species in chloroplasts, one of 5S, and a 4S type that appears to be derived, with the 23S, by evolution from the bacterial 23S rRNA. There are 21 peptides in the small subunit of both bacterial and chloroplast ribosomes, but the large subunits differ slightly (32 in chloroplasts, 31 in *E. coli*).

Close to the 5′ end of the mRNA, there is a 'start codon' (codon: a group of three bases) to which the ribosome attaches. Amino acids are meanwhile attached to molecules of tRNA; each type of tRNA has a specific 'anticodon' that will recognise and pair with the appropriate codon when it is presented at the ribosome site, where the amino acid is added to the growing peptide chain. In this way, counting in threes towards the 3′ end, each successive amino acid of the intended protein (reading from the N- to the C-termini) is added as dictated by the codon sequence until a 'stop' codon is reached. The correspondence of codons with amino acids and stop–start signals is the 'genetic code'. The code is said to be 'degenerate', meaning that there is usually more than one codon for a given amino acid.

In chloroplasts and bacteria, but not notably in nuclei, the primary transcript of RNA is often polycistronic, that is, one mRNA molecule may carry code for several proteins. Polycistronic proteins are regulated together by a single promoter sequence, and are said to be co-transcribed.

There is no structural division of the mRNA or DNA into codons; translation can begin at any set of three bases that reads the start codon, and thereafter continues in threes. It is therefore possible, although rare, for genes to partially overlap.

The translating activity of ribosomes can be inhibited: 70S ribosomes by antibiotics such as chloramphenicol, and 82S ribosomes by cycloheximide. This provides a simple if crude test for the site of synthesis of a particular protein: if it is synthesised in the cytoplasm, its production will be inhibited by cycloheximide, and if it is synthesised in the chloroplast, by chloramphenicol.

13.4 General Principles of Gene Isolation

13.4.1 Restriction endonucleases

The DNA of the genome must be fragmented in a controlled manner so that the fragments are small enough to work with, but without losing the ability to relate them in the right order to the original whole circle or chromosome. This can be done by means of restriction endonucleases. These are enzymes that are obtained from various bacteria, and which have the property of recognising specific sequences of bases, usually four or six, and cleaving the DNA (both strands) in a particular way (see Fig. 13.2) by hydrolysis.

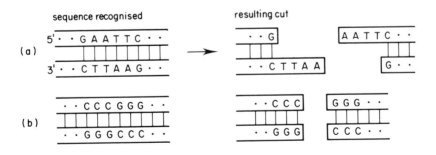

Figure 13.2.   Restriction endonucleases: recognition sites. The two enzymes Eco R1 (a) and Sma 1 (b) are given as examples: they recognise the specific DNA sequences shown, and cleave them so as to leave 'sticky ends' and 'blunt ends' respectively. Note that the sequences are self-pairing

A restriction enzyme that looks for a six-base sequence will find it on average once in every 4096 bases (4^6), since the sequence of bases appears to be random. Most recognition sequences are self-pairing (the same sequence in each strand, see Fig. 13.2). (The chloroplast genome could yield on this basis 37 fragments, averaging 4096 base pairs each, without allowing for the fact that in chloroplast DNA A+T pairs are more abundant than G+C pairs.) The fragments produced by the restriction enzyme can be separated by electrophoresis on agarose gels, and either be collected by elution or 'blotted' onto an adsorbant surface for testing.

Digests of a chloroplast genome with different restriction enzymes yield different sets of fragments. Because the recognition sites occur virtually at random, the sets overlap; a fragment obtained with one enzyme can be tested for the presence of recognition sites for other enzymes. In this way the fragments can be arranged in their correct order. The restriction sites are the points of reference when mapping genes on a DNA molecule.

The fragments of DNA blotted on a surface can be tested for their relationship to rRNA or tRNA by first heating the surface to denature the double-stranded structure into single strands, and allowing it to cool in the presence of ^{32}P-labelled rRNA or tRNA. The DNA pairs with an RNA molecule if the sequences are complementary; this is termed hybridisation. The unhybridised RNA is washed away, and the hybridised DNA fragments identified by the radioactivity of the attached RNA. In this way the segments of DNA that code for these two forms of RNA can be recognised and mapped.

The identification of the DNA segments that contain code for mRNA (and hence for protein) is more difficult, because there are many more different species of mRNA than of rRNA or tRNA, and it is impractical to attempt to purify even one. In most cases the experimenter starts with a known protein which has been purified to the point where either an antibody can be obtained, or some of the amino acid sequence can be established (say 10 amino acids). The cells that manufacture that protein are expected to contain the mRNA for it, and the total mRNA is extracted from the cells. Without separating the many kinds (thousands) of mRNA, the researcher uses the enzyme RNA-dependent DNA polymerase to form a version of DNA, known as copy-DNA (cDNA), for each one (see Fig. 13.3). The mixture of cDNA is amplified and tested as described below.

Alternatively, one can study fragments of the actual genome. Genomes of viruses, chloroplasts and mitochondria can be completely sequenced, but larger ones such as nuclear chromosomes are too big, and it is normal practice to make 'libraries' by fragmenting the genome, amplifying the fragments (see below), and resolving them only when it is deemed profitable to make a search. With a cDNA collection, every cDNA is representative of a protein, but with a genomic library, most of the DNA may not be known to code for anything, and the restriction cuts are not likely to coincide with gene boundaries.

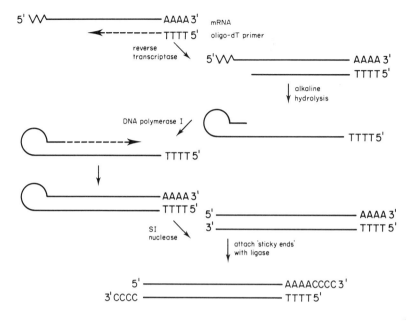

Figure 13.3. Stages in the production of cDNA. The AAAA sequence represents the poly-A tail on mRNA

13.4.2 Amplification of restriction fragments of cDNA

The purpose of amplification is to multiply the quantity of each DNA fragment in the mixture until there is sufficient for testing and analysis.

The method involves the use of a vector, derived from a bacterial plasmid. Plasmids are circular molecules of double-stranded DNA that grow, under suitable conditions, in bacterial cells, in some ways like parasites, and they can be used to produce large quantities of gene fragments or cDNA molecules. A suitable plasmid is grown in a host culture of bacteria, and collected after lysis of the cells. The DNA circles are opened by means of a restriction enzyme, and it is arranged that the ends of the gene fragments, or the cDNA molecules, match the ends of the plasmid DNA. The opened plasmid preparation is mixed with the mixture of gene fragments or cDNA molecules, and the DNA is then joined up using a ligase enzyme so that the plasmid circle is re-formed, but is now somewhat larger because of the included DNA sequences. The experimenter relies on the statistical likelihood that each cDNA or genome fragment will be contained in at least one plasmid. Next, a culture of bacteria is infected with the plasmid preparation, and grown through several divisions so that each original cell, and the plasmids in it, gives rise to many identical progeny. A sample is plated out. It is possible to recognise colonies that contain plasmids, and those that contain plasmids that carry fragments of the DNA under investigation. The colonies of interest are preserved and grown in bulk.

13.4.3 Matching DNA fragments to proteins

If the (partial) amino acid sequence of the target protein is known, an artificial DNA (probe-DNA) can be synthesised according to the genetic code, guessing which codon to use where the code is degenerate. About 30 bases are assembled, to correspond with a known sequence of ten amino acids. The probe-DNA is made radioactive. Each agar plateful of colonies is printed by contact onto a surface (usually cellulose nitrate), and the cells that adhere to it are lysed by heating so that the DNA is left adhering in a single-stranded form to the surface. The entire surface is irrigated with the probe-DNA. Radioactive spots remaining after washing are interpreted as hybrids between the probe and the cell DNA, and the corresponding colony on the original plate is cultured. The plasmids are re-isolated from the cells, and cleaved with the same restriction enzyme so as to cut out the cDNA or genomic fragment in milligram quantities.

In the procedure described above, the plasmid vector amplified the DNA fragment. It is alternatively possible for some vectors to cause the host bacterial cell to transcribe and translate the fragment, and this approach is useful if an antibody to the target protein is available. The colonies of bacteria would be tested with the antibody (with an enzyme label attached), thus identifying which of them had synthesised the target protein, and hence which contained the cDNA corresponding to it.

Genomic fragments are not necessarily genes, and the above procedure identifies those fragments which contain sequences that are close to the sequence of the probe-DNA. The actual gene is found by determining the sequence of the genomic fragment that contains the probe sequence, and inspecting it for start and stop codons. This is reasonably practicable for chloroplasts and bacteria in which introns are rare, but it may be impossible with nuclear genes.

It has happened (notably in the chloroplast) that nucleotide sequencing of genomic fragments has revealed open reading frames (meaning translatable sequences of DNA of reasonable size with start and stop codons) for which there was no target protein. Such sequences are not necessarily expressible genes (for example they may lack a promoter), but if they can be matched with the sequences of known genes (in other species) or with amino acid sequences of known proteins, there is a prima facie case for considering them as genes. Several such sequences have been discovered: for example, in pea and wheat chloroplast DNA an open reading frame has been found (atpI) which matches the sequence of subunit a of the F_0 particle of the *E. coli* coupling factor (see (146) for an example).

Once one gene has been identified in one species, the homologous gene can be located in related species. The known gene can be broken down into small pieces and the pieces tested for hybridisation to a genomic library from the target species. If the gene fragments are small and sufficiently numerous to avoid introns, they are likely to hybridise with the library fragments that belong to the gene of the target species.

13.5 DNA Sequencing

The sequence of the bases in a fragment of DNA can be determined by a method developed by F. Sanger and honoured by the Nobel Prize in 1978. Essentially, single-stranded DNA is used as a template in a system containing DNA polymerase which assembles its complementary strand from a mixture of deoxynucleoside triphosphates. Each nucleotide is added by the enzyme so that the phosphate attached to the 5'-hydroxyl of its 2'-deoxyribose forms an ester bond with the 3'-hydroxyl at the end of the growing DNA strand. In each of four parallel incubation mixtures, a radioactive, di-deoxy nucleoside triphosphate (ddATP, ddGTP, ddCTP or ddTTp) is added which lacks the 3'-hydroxyl group on the ribose; it will be incorporated at random by substituting for the appropriate proper nucleotide (dATP, dGTP, dCTP or dTTP respectively). Once the dd-nucleotide has been incorporated, the strand cannot grow further because it has no 3'-hydroxyl group for the next nucleotide to link to. At the end of the incubation, the duplexes are separated, and each sample will contain DNA strands of different lengths, each strand terminating in a radioactive dd-nucleotide that corresponds to one particular proper base. Electrophoresis separates the samples in parallel tracks, each track corresponding to the set of strands that have stopped growing after the incorporation of the dd-version of A, G, C or T. Each base in the sequence corresponds to a radioactive spot in the track in which that base was represented in the dd-nucleotide, and each spot has travelled a different distance, proportional to its length. No two strands have the same length. From the autoradiograph, the sequence can be read, several hundred bases at a time (see Fig. 13.4).

If the DNA under examination was a genomic fragment, the sequence would be scanned for start and stop codons, so as to detect open reading frames. Using the genetic code, amino acid sequences are then written down corresponding to the sequence of codons in the reading frame. The sequence is then compared with libraries of known protein sequences, stored in computers. Often proteins belong to families, with homologies in their sequences, and this aids such a library search.

13.6 Genomic Maps of Higher-plant Chloroplasts

The complete nucleotide sequences of chloroplast DNA have been established for two species: the liverwort *Marchantia polymorpha* (235), and tobacco *Nico-*

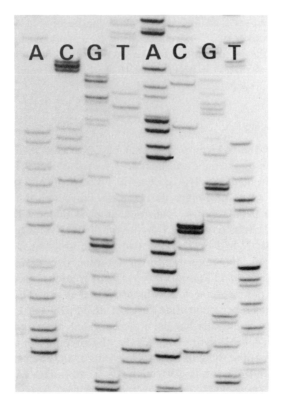

Figure 13.4. Autoradiograph of DNA sequences, all beginning at the same point but terminating in a radiolabelled-dideoxynucleotide. The sequence of the parent DNA can be read as shown. Autoradiograph, of yeast DNA, courtesy of Dr S. Smith

tiana tabacum (236). The liverwort chloroplast genome (121,024 base pairs) is smaller than that of tobacco (155,844 bp), mainly because of the different lengths of the inverted repeat sequence (see Fig. 13.5). Ohyama et al, who determined the liverwort sequence, in a stimulating article (237) point out that in spite of the difference in size, the total numbers of genes represented are very similar, 119 (plus 9 duplicates) in the liverwort, and 122 plus 24 duplicates in tobacco. Apart from a group of leguminous plants (notably pea), all chloroplast DNAs have an inverted repeat and hence can be divided (in the imagination) into four regions: inverted repeats IRA and IRB, and short and long single copy sequences SSC and LSC. The largest group of genes (about 60) are concerned with the transcription–translation system. Genes for rRNAs are found in the IR sequences, and are transcribed as a single RNA of mass 2.7MDa. This primary transcript is cleaved to provide the 4.5S, 5S, 16S and 23S rRNAs, and also four tRNAs. The tRNA genes (32 plus 5 repeats) are apparently distributed at random.

The chloroplast genome resembles that of prokaryotes in that several groups of genes are co-transcribed to produce an mRNA coding for two or more proteins. An example is given by the *psb*C and *psb*D genes, which are contiguous in Fig. 13.5 and actually overlap by 50 base pairs in the wheat genome (146). It has been shown by a technique known as 'Northern blotting' (in which RNA molecules, separated by electrophoresis, are tested with a probe-DNA) that the open reading frame ORF62 is included in the transcript, and could be generating a polypeptide of 62 amino acids, not yet identified. The presumed amino acid sequence could include two membrane-spanning segments. Will it be a 6 kDa membrane protein?

The chloroplast genome differs from the usual prokaryote genome in that several genes contain introns (for example the *pet*B and *pet*D genes). Introns are known in mitochondrial DNA, however, and particularly in nuclear DNA, where they are many times longer in total than the exons. Ohyama et al (235, 237) drew attention to the 30-odd open reading frames, which were conserved in the liverwort and tobacco genomes, and to the newly discovered (chloroplast) gene groups *ndh*, *frx* and *mbp*. The significance of these is discussed in the next section. A few genes are present in one but not the other genome, suggesting that during evolution a few genes have been rearranged, and interchanged with the nuclear genome.

Nineteen genes code for known membrane proteins. These are listed in Tables 13.1–13.5. In most cases the complexes in the thylakoid membrane are composed of both nuclear and chloroplast-coded polypeptides, but the entire PSII intrinsic complex (apart from possible small polypeptides) is coded by the eight genes *psb*A–H.

Recently (235, 237) the complete sequencing of the chloroplast genome (of a liverwort) detected a number of open reading frames (see Fig. 13.5) which were matched to known genes in human mitochondria. They code for polypeptide components of the NADH : ubiquinone reductase (complex I). These ORFs, labelled *ndh*1–6, are actively transcribed and provide a rationale for a chloroplast respiratory electron transport system reported in *Chlamydomonas* chloroplasts (257).

The liverwort and tobacco chloroplast genomes contain *frx* genes, of which *frx*A was shown to produce the 8 kDa polypeptide of PSI. The sequence corresponds closely with bacterial ferredoxin (Fe_4S_4), confirming the existence of such an iron–sulphur centre in PSI. The gene *frx*B is very similar to *frx*A but its product is not known. The liverwort also contains the gene *frx*C, which is related to part of the nitrogen-fixing apparatus of cyanobacteria. It has a section of its sequence that is characteristic of nucleotide-binding proteins. The same sequence occurs in the *mbp* genes (237) which are related to ATP-driven amino acid permeases in the inner membrane of bacteria.

The only nuclear coded peptides in PSII are the 32, 24 and 16 kDa extrinsic peptides that protect the oxygen-evolving system (see Table 13.2). These are synthesised on cytoplasmic ribosomes, and therefore have to pass through three membranes (the double envelope and the thylakoid membrane) in order to reach their destination in the thylakoid lumen. It may be that the leading peptide at the N-terminus is a 'transit peptide' for passage through the envelope, and is hydrolysed off in the stroma, while the trailing peptide at the C-terminus is a transit peptide for passage through the thylakoid. After removal of the trailing peptide, the mature protein assembles onto PSII on the luminal side of the membrane.

Two genes (*psa*A, B) (Fig. 13.5) have been mapped that code for the major peptide of PSI (CCI) and another (*frx*A) has been identified with an Fe_4S_4 polypeptide and renamed *psa*C (see Table 13.1). Other (minor) PSI peptides are known to be nuclear coded.

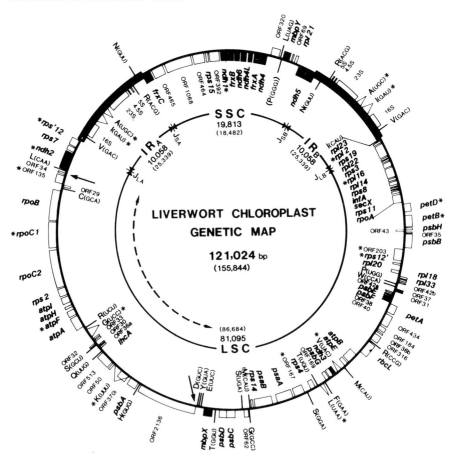

Table 13.1 Photosystem I protein components.

Component			Mr	Mc*	Gene and location	
Core proteins						
Subunit	IA	Reaction	66–67K	83.2K	*psa*A	(cp)
	IB	centre		82.5	*psa*B	(cp)
	II		21			(n)
	III		17			(n)
	IV		13			(n)
	V		11			(n)
	VI		9			(n)
EA (electron acceptor) protein			8	8.9	*psa*C (*frx*A)†	(cp)
Antenna proteins						
LHCI			20,22,23	24	*Cab*	(n)

* molecular weight calculated from coding sequence
† initial designation

Tables 13.1–13.5 were kindly supplied by Dr Tristan A. Dyer.

Figure 13.5. Genetic map of the completely-sequenced chloroplast genome from a liverwort, *Marchantia* compared with that of tobacco. The inner circle, SSC, IRB, LSC, IRA gives a summary of the general plan, in which the four regions are small single copy, inverted-repeat B, large single copy and inverted-repeat A, correspondingly. The numbers are the number of base-pairs in each region of the liverwort chloroplast genome, and the parentheses contain the corresponding figures for tobacco. The dashed segment drawn inside the inner circle and the arrows outside it show the region of an inversion with respect to tobacco. Genes are labelled with abbreviations according to Ohyama et al (235) and Crouse et al (256). Asterisks indicate genes containing introns. Genes shown on the outside of the diagram are transcribed anticlockwise. The symbol ORF 2136 (etc.) denotes an open reading frame that could code for a hypothetical protein containing 2136 amino acids. Some open reading frames have been allocated gene-symbols based on homologies in their sequences: *ndh*1–5, homologous to mitochondrial subunits of complex I; *frx*A–C, homologous to Fe_4S_4 proteins elsewhere; *mbp*X, homologous to ATP-binding subunit of a bacterial permease and *lhc*A, homologous to a bacterial light-harvesting protein. The symbol A(UGC) indicates the gene for the transfer RNA for alanine with UGC as its anticodon. Solid boxes represent new genes reported in Ohyama et al (237). Reproduced by permission of Professor K. Ohyama and Elsevier Publications Cambridge. From Ohyama et al (1988) *Trends Biochem. Sci.* **13**: 20

Table 13.2   Photosystem II (PSII) protein components.

Component			Mr	Mc*	Gene and location	
Core proteins						
DI	}	Reaction	32K	$\approx 37.2K$	*psb*A	(cp)
D2		centre	32	39.4	*psb*D	(cp)
Cytochrome b-559 (α & β)			10	9.4,4.4	*psb*E,F	(cp)
Antenna proteins						
Chlorophyll *a*-binding			47	56.2	*psb*B	(cp)
antenna proteins			44	51.8	*psb*C	(cp)
LHCII	Type 1	} Chl *a/b*	25–28	25.2	*Cab*	(n)
LHCII	Type 2	binding			*Cab*	(n)
CP29		protein	29			
Oxygen-evolving complex						
OE1			33	26.5	*wox*A	(n)
OE2			23	20.2	*wox*B	(n)
OE3			16	16.5	*wox*C	(n)
Others						
Phosphoprotein			10	7.8	*psb*H	(cp)
Intrinsic proteins			24			(n)
			22			(n)
			10			(n)

* molecular weight calculated from coding sequence

Table 13.3   Cytochrome b_6f protein components.

Component	Mr	Mc*	Gene and Location	
Cytochrome *f* apoprotein	31.3 (41)	31.9 (35.3)†	*pet*A	(cp)
Cytochrome b-563	23	23.7	*pet*B	(cp)
Rieske protein	18–20	18.8 (26.8)†		(n)
Subunit IV	15.2–17.5	15.2	*pet*D	(cp)
Subunit V	19.5			(n)

* molecular weight calculated from coding sequence
† precursor sizes

The *bf* complex is chloroplast coded (genes *pet*A, B and D) (Fig. 13.5 and Table 13.3) except for the Rieske iron–sulphur protein. The cytochrome b_6 gene (*pet*B) corresponds to the N-terminal sequence of mitochondrial cytochrome *b*, and the chloroplast gene *pet*D to the C-terminal sequence of mitochondrial *b*. Introns are suspected to exist in both. The two genes are co-transcribed in spinach. The *pet*A gene is widely separated from the rest in chloroplasts, but in the bacteria *Rhodopseudomonas* and *E. coli* they are all transcribed on one operon.

The ATPase particle (CF, see Chapter 3) has six chloroplast genes (*atp*A, B, E, F, H, I), three for CF_1 and three for CF_0. The nucleus contributes two and one genes respectively (Table 13.4). The three adjacent genes *atp*H, F and A are co-transcribed, and *atp*I is an assumption based on an open reading frame that precedes *atp*H, and is homologous with the sequence of the subunit F_0-a in *E. coli*.

Table 13.4 ATP synthase components.

Component			Mr	Mc*	Gene and Location	
F_1	α	catalytic	58K	55.4K	*atp*A	(cp)
	β		57	53.9	*atp*B	(cp)
	γ	S-S regulation	38	–	*atp*C	(n)
	δ	F_0 link	25	–	*atp*D	(n)
	ε	regulation	14	14.7	*atp*E	(cp)
Fo	a(IV)	proton pore	19	20 (27.1)†	*atp*I	(cp)
	I	F_0 anchor	18	19 (20.9)†	*atp*F	(cp)
	II		16	–	*atp*G	(n)
	III	proton-translocating	8	8	*atp*H	(cp)

* molecular weight calculated from coding sequence
† precursor sizes
F_1 stoichiometry $\alpha_3\ \beta_3\ \gamma_1\ \delta_1\ \varepsilon_1$
F_0 stoichiometry $a_1\ I_1\ II_1\ III_{10}$

The three soluble proteins, plastocyanin, ferredoxin and FNR, are all nuclear coded, as is the small subunit of the enzyme RuBisCO. The large subunit is coded in the chloroplast, and was the first identified chloroplast protein gene product. Under normal conditions the synthesis is coordinated so that the small subunit is synthesised (on cytoplasmic ribosomes from nuclear mRNA) at the same rate that the chloroplast synthesises the large subunit. If synthesis of the large subunit is blocked, the small subunit is synthesised but then rapidly broken down in the cytoplasm; it is neither transported nor accumulated. The small subunit, which also has a leading transit sequence, enters by means of a specific recognition protein in the chloroplast envelope (219) (but see Joyard and Douce (246)). There is no obvious homology between the transit peptides of RuBisCO from pea and *Chlamydomonas*, so it is unlikely that the sequence itself provides a mechanism for crossing the membrane. It is possible that the transport of the small subunit, and the regulation of synthesis of the large subunit, may be related, and further information on the synthesis and the coding of the receptor protein might clear the matter up.

In cyanobacteria the genes for the large and small subunits of RuBisCO are adjacent on the same genome (the only genome) and they are also found on the same (chloroplast) genome in two groups of algae, Chromophyta and Rhodophyta. This supports the idea (see Chapter 2) that these groups are relatively primitive and close to an ancestral cyanobacterial form.

13.7 The Impact of Molecular Genetics on the Study of Photosynthesis

13.7.1 Mechanisms

We study structures so as to infer mechanisms, and the rapidity of gene sequencing, and of comparative gene sequencing (of the same gene in different organisms), is such an improvement on protein sequencing by traditional methods that we find insights into (a) the roles of membrane-spanning segments in intrinsic membrane proteins, (b) the ligand groups for iron–sulphur and copper proteins, and (c) homologies identifying unexpected peptides.

By way of an example for (a), studies of the sequences of the membrane-spanning segments of chlorophyll-carrying proteins, and comparisons between different proteins and between different species, reveal that certain groupings are highly conserved (that is, they are the same in many species representing different evolutionary stages), and are therefore likely to be essential for functioning. Such groups are the histidine residues that are thought to be the point of the (coordinate) attachment of chlorophyll molecules, and the groups of cysteine residues that are required for binding iron–sulphur clusters. By this means the *psa*A and B genes, coding for the PSI major peptides, are confirmed as the carriers of both chlorophyll and iron–sulphur clusters (see Chapter 6).

For an example of (b), the genes *psb*E and F (which were identified in a genomic library by means of the antibody against purified cytochrome b-559) showed the presence of one histidine residue in each peptide (9 and 4 kDa respectively). It could therefore be suggested that the haem iron bridged between them.

An impressive example of (c) was the discovery that the sequences of the two peptides known as the A and B proteins (D_2 and D_1 respectively) in PSII (238, 239) were very similar to the sequences for the M and L subunits of the purple bacterial reaction centre (33), completely transforming our thinking on the nature of the PSII reaction-centre, which had become somewhat stalled. A second example was the matching of mitochondrial genes for complex I to ORFs in the chloroplast genome (237, see p. 219).

Table 13.5 Genes for putative protein components of a chloroplast NAD(P)H quinone oxidoreductase.

Chloroplasts		Equivalent mitochondrial component	
Marchantia	Tobacco	Gene	Mc
*ndh*1	*ndh*A	*ND*1	24
2	B	2	25
3	C	3	6
4	D	4	36–39
4L	E	4L	3.5
5	F	5	51
6	G*	6	18

* ORF 138 + ORF 99B (see Ohyama et al (237))

13.7.2 Recognition

Proteins interact; they recognise each other by means of features that are not always obvious. The technique of site-directed mutagenesis enables the researcher to produce virtually unlimited numbers of variations on a protein, so that their relative efficiency can, in principle, be deconvoluted to reveal the influence of their primary structure on recognition and, say, catalytic activity.

Site-directed mutagenesis is a technique used to investigate whether or not a particular amino acid is essential for the functioning of a protein. Fragments of the gene are prepared, and the fragment containing the codon for the target amino acid selected. A section of DNA is synthesised containing the altered codon, and the gene reassembled, incorporating the modified segment, by means of ligases. At present, there is no means of re-inserting genes into the chloroplast genome, still less of getting them to be expressed in the chloroplast; the multiple copies make the proposition impracticable. However, altered chloroplast genes can be expressed in microbial systems, allowing the preparation of mutant protein for biochemical testing.

With bacteria, and cyanobacteria, means exist to transfer DNA from outside the cells into the genome (transformation). In this way site-directed mutagenesis in the *psb*D gene was used to remove the specific histidines from the D_2 peptide in PSII of *Synechocystis*. The effect was a loss of PSII activity. This gave support to the theory that the histidines were needed for binding components such as P680 and the Q_A–Fe complex (but the authors, Vermaas et al (247), pointed out that electron transport can be inhibited by mutations of antenna complexes, and advised caution in reaching too rapid a conclusion!).

The same gene was attacked by the same method in order to remove tyrosine–

160 from the D_2 polypeptide. The effect was to remove ESR signal II_u, ascribed to the second donor D in PSII (see Section 5.5.2), and supporting the hypothesis that Z and D were the corresponding tyrosines in the D_1 and D_2 polypeptides respectively (245).

A field which is likely to produce examples in the near future is the electron-transport sequence between the bc_1 particle and the reaction centre of *Rps capsulata*, mediated by cytochrome c_2. In several laboratories there now exist the cloned genes for cytochrome c_2 (*cyc*A), the *puf* operon for the reaction centre M and L subunits, and the *pet* operon for the *bc* complex. These genes are all now freely manipulable.

13.7.3 Control

The genome is regulated so that different genes (or operons) are turned on at different stages of the life-cycle, or in response to environmental conditions.

For example, the *puf* operon in *Rps capsulata* contains the codes for the M and L subunits of the reaction centre, and also the α- and β-subunits of the antenna protein, B870. This operon is induced (switched on) by a fall in the concentration of oxygen during aerobic culture, or a fall in the light level during photosynthetic culture. Control is exercised at the transcriptional level. The same is true of the enzymes that synthesise the pigments; protein and pigment synthesis are co-regulated. It has been suggested that oxygen tension is mediated by the effect on the oxidase that converts the photosynthetic pigment spheroidene to the aerobic component spheroidenone, although the interrelationships are complex.

Figure 13.1 showed the promoter region, at -10 bases upstream, known as the 'TATA box'. Longer sequences (-330 to -50 bases upstream) have been found for some nuclear genes. The 'LRE' sequences are the sites at which phytochrome (indirectly) exerts control as a 'light-switch' (258).

13.8 Limitations

Proteins that carry prosthetic groups such as chlorophyll, acting as antenna or reaction-centre complexes, achieve an accuracy of spacing and orientation that reduces energy loss to spectacularly low levels. In principle the folding of the peptide and the binding of chlorophyll could be predicted from minimum-energy calculations based on the primary sequence (provided it was known what degree of post-translational proteolysis had taken place up to the point at which the final conformation and pigment binding took place), but the achievement of the accuracy required seems many years away, and without such prediction amino acid sequences tell us little about the essential operation of pigmented proteins.

The molecular geneticist is at present most strongly placed to discuss family and evolutionary relationships, and is working to establish the features of DNA organisation that lead to the differential expression of genes during the development of an individual. These topics are not directly related to photosynthesis, but indirectly, they are an encouragement to biochemists to view an organism as a complete lifetime entity.

Further Reading

Useful introductions to the techniques and reasoning of molecular biology (meaning the manipulation of DNA and RNA) are:

Grierson, D. & Covey, S. (1984) *Plant Molecular Biology*. Tertiary Biology Series, Blackie, Glasgow.

Mainwaring, W.I.P., Parish, J.H., Pickering, J.D. & Mann, N.H. (1982) *Nucleic Acid Biochemistry and Molecular Biology*, Blackwell Scientific Publications, Oxford.

References

1 Staehelin, L.A., Armond, P.A. & Miller, K.R. (1976) Chloroplast membrane organisation at the supramolecular level and its functional implications. *Brookhaven Symp. Biol.* **28**: 278–315.
2 Staehelin, L.A. (1986) Chloroplast structure and supramolecular organization in photosynthetic membranes. In Staehelin, L.A. & Arntzen, C.J., eds, *Encyclopaedia of Plant Physiol.*, new series, vol. 19, Springer-Verlag, Berlin, pp. 1–84.
3 Andersson, B. & Anderson, J.M. (1980) Lateral heterogeneity in the distribution of the chlorophyll–protein complexes of the thylakoid membranes of spinach chloroplasts. *Biochim. Biophys. Acta* **593**: 427–440.
4 Smith, J.H. & Benitez, A. (1955) Chlorophylls: analysis in plant materials. *Modern Methods of Plant Analysis*, vol. 4, Springer-Verlag, Berlin, pp. 142–196.
5 Anderson, J.M. & Barrett, J. (1986) Light-harvesting pigment–protein complexes of algae. In Staehelin, L.A. & Arntzen, C.J., eds, *Encyclopaedia of Plant Physiol.*, new series, vol. 19, Springer-Verlag, Berlin, 269-285.
6 Scheer, H. (1986) Excitation transfer in Phycobilins. In Staehelin, L.A. & Arntzen, C.J., eds, Photosynthesis III. *Encyclopaedia of Plant Physiol.*, new series, vol. 19, Springer-Verlag, Berlin, pp. 327–337.
7 Zuber, H. (1986) Structure of light-harvesting antenna complexes of photosynthetic bacteria, cyanobacteria and red algae. *Trends Biochem. Sci.* **11**: 414–419.
8 Feick, R.G. & Fuller, R.C. (1984) Topography of the photosynthetic apparatus of *Chloroflexus aurantiacus*. *Biochemistry* **23**: 3693–3700.
9 Fenna, R.E. & Matthews, B.W. (1977) Structure of a bacteriochlorophyll a-protein from *Prosthecochloris aestuarii*. In Olson, J.M. and Hind, G., eds, *Chlorophyll-Proteins, Reaction Centers and Photosynthetic Membranes*, Brookhaven Symp. Biol., **28**, pp. 170–182.
10 Zuber, H. (1986) Primary structure and function of the light harvesting polypeptides from cyanobacteria, red algae, and purple photosynthetic bacteria. In Staehelin, L.A. & Arntzen, C.J., eds, Photosynthesis III. *Encyclopaedia of Plant Physiol.*, new series, vol. 19, Springer-Verlag, Berlin, pp. 238–251.
11 Thornber, J.P., Trosper, T.L. & Strouse, C.E. (1978) Bacteriochlorophyll in vivo: relationship of spectral forms to specific membrane components. In Clayton, R.K. & Sistrom, W.R., eds, *The Photosynthetic Bacteria*. Plenum, New York, pp. 133–160.
12 Ogawa, T., Obata, F. & Shibata, K. (1966) Two pigment-proteins in spinach chloroplasts. *Biochim. Biophys. Acta* **112**: 223–224.
13 Thornber, J.P., Smith, C.A. & Bailey, J.L. (1966) Partial characterization of two chlorophyll–protein complexes isolated from spinach beet chloroplasts. *Biochem. J.* **100**: 14–15.
14 Thornber, J.P. (1986) Biochemical characterization and structure of pigment-proteins of photosynthetic organisms. In Staehelin, L.A. & Arntzen, C.J., eds, *Encyclopaedia of Plant Physiol.*, new series, vol. 19, Springer-Verlag, Berlin, pp. 98–142.
15 Kuhlbrandt, W. (1984) Three-dimensional structure of the light-harvesting chlorophyll a/b protein complex. *Nature* **307**: 478–480.

16 Kyte, J. & Doolittle, R. (1982) A simple method for displaying the hydropathic character of a protein. *J. Mol. Biol.* **157**: 105–132.

17 Bennett, J., Steinback, K.E. & Arntzen, C.J. (1980) Chloroplast phosphoproteins: regulation of excitation energy transfer by phosphorylation of thylakoid membrane peptides. *Proc. Natl Acad. Sci. USA* **77**: 5253–5257.

18 Allen, J.F., Bennett, J., Steinback, K.E. & Arntzen, C.J. (1981) Chloroplast protein phosphorylation couples plastoquinone redox states to distribution of excitation energy between photosystems. *Nature* **291**: 21–25.

19 Haworth, P. (1983) Protein phosphorylation-induced State I–State II transitions are dependent on thylakoid membrane viscosity. *Arch. Biochem. Biophys.* **226**: 145–154.

20 Allen, J.F., Harrison, M.A., Holmes, N.G., Mullineux, C.W. & Sanders, C.E. (1987) Regulation of photosynthetic unit function by protein phosphorylation. In Biggins, J., ed., *Progress in Photosynthesis Research*, Proc. VIIth Int. Congr. Photosynth. Res., 1986, vol. 2, pp. 757–760.

21 Bennett, J. (1977) Phosphorylation of chloroplast membrane polypeptides. *Nature* **269**: 344–346.

22 Camm, E.L. & Green, B.R. (1981) A model of the relationship of the chlorophyll–protein complexes associated with photosystem II. In Akoyunoglou, G., ed., *Photosynthesis*, Proc. Vth Int. Congr. Photosynth. Res. 1980, vol. 3, pp. 675–681.

23 Burger-Wiersma, T., Veenhuis, M., Korthals, H.J., Van de Wiel, C.C.M. and Mur, L.R. (1986). A new prokaryote containing chlorophylls a and b. *Nature* **320**: 262–264.

24 Mullet, J.E., Burke, J.J. & Arntzen, C.J. (1980) Chlorophyll proteins of photosystem I. *Plant Physiol.* **65**: 8814–8822.

25 Haworth, P., Watson, J.L. & Arntzen C.J. (1983) The detection, isolation and characterization of a light-harvesting complex which is specifically associated with photosystem I. *Biochim. Biophys. Acta* **724**: 151–158.

26 Lam, E., Ortiz, W. & Malkin, R. (1984) Chlorophyll a/b proteins of photosystem I. *FEBS Lett.* **168**: 10–14.

27 Bassi, R. & Simpson, D. (1987) Chlorophyll–protein complexes of barley photosystem I. *Eur. J. Biochem.* **163**: 221–230.

28 White, M.J. & Green, B.R. (1987) Antibodies to the photosystem I chlorophyll a + b antenna cross-react with polypeptides of CP29 and LHCII. *Eur. J. Biochem.* **163**: 545–551.

29 Olive, A., Wollman, F.A., Bennoun, P. & Recouvreur, M. (1983) Localization of the core and peripheral antennae of photosystem I in the thylakoid membranes of *Chlamydomonas reinhardtii*. *Biol. Cell.* 81–84.

30 Reed, D.W. & Clayton, R.K. (1968) Isolation of a reaction-center from *Rhodopseudomonas sphaeroides. Biochem. Biophys. Res. Comm.* **30**: 471–475.

31 Gingras, G. & Jolchine, G. (1969) Isolation of a P870-enriched particle from *Rhodospirillum rubrum.* In Metzner, H., ed., *Progress in Photosynthesis Research*, Proc. Int. Congr. Photosynth. Res. 1968, pp. 209–216.

32 Thornber, J.P., Olson, J.M., Williams, D.M. & Clayton, M.L. (1969) *Biochim. Biophys. Acta* **172**: 351–354.

33 Hearst, J.E. & Sauer, K. (1984) Protein sequence homologies between portions of the L and M subunits of the reaction centers of *Rhodopseudomonas capsulata* and the 32 kD herbicide-binding polypeptide of chloroplast thylakoid membranes and a proposed relation to quinone binding sites. *Z. Naturforsch.* **39c**: 421–424.

34 Deisenhofer, J., Epp, O., Miki, K., Huber, R. & Michel, H. (1984) X-ray structure analysis of a membrane protein complex. *J. Mol. Biol.* **180**: 385–398.

35 Hurt, E.C. & Hauska, G. (1984) Purification of membrane-bound cytochromes and a photoactive P-840 photoactive protein complex of the green sulphur bacterium *Chlorobium limicola* f. *thiosulfatophilum. FEBS Lett.* **168**: 149–154.

36 Green, B.R. & Camm, E.L. (1984) Evidence that CP47 (CPa-1) is the reaction center of Photosystem II. In Sybesma, C., ed., *Advances in Photosynthesis Research*, Proc. VIth Int. Congr. Photosynth. 1983, Nijhoff, The Hague, vol. 2, pp. 95–98.

37 Barber, J. (1987) Photosynthetic reaction centres: a common link. *Trends Biochem. Sci.* **12**: 321–326.

38 Satoh, K. & Nanba, O. (1987) Isolation of a photosystem II reaction center consisting of gamma and delta subunits (D-1 and D-2) and cytochrome b-559. In Biggins, J., ed., *Progress in Photosynthesis Research*, Proc. VIIth Int. Congr. Photosynth. 1986, Nijhoff, Dordrecht, vol 2, pp. 69–72.

39 Lundell, D.J., Glazer, A.N., Melis, A. & Malkin, R. (1985) Characterization of a cyanobacterial photosystem I complex. *J. Biol. Chem.* **260**: 646–654.

40 Wolff, Ch. & Witt, H.T. (1971) On metastable states of carotenoids in primary events of photosynthesis. In Forti, G., Avron, M. & Melandri, N., eds, *Photosynthesis Two Centuries after its Discovery by Joseph Priestley*, Proc. IInd Int. Congr. Photosynth. Res. 1971, The Hague, pp. 931–936.

41 Krasnowskii, A.A. (1969) The principles of light energy conversion in photosynthesis: photochemistry of chlorophyll and the state of pigments in organisms. In Metzner, H., ed., *Progress in Photosynthesis Research*, Proc. Int. Congr. Photosynth. Res. 1968, Metzner, Tubingen, pp. 709–727.

42 French, C.S. & Prager, L. (1969) Absorption spectra for different forms of chlorophyll. In Metzner, H., ed., *Progress in Photosynthesis Research*, Proc. Int. Congr. Photosynth. Res. 1968, Metzner, Tubingen, pp. 555–564.

43 French, C.S. & Huang, H.S. (1957) The shape of the red absorption band of chlorophyll in live cells. *Carnegie Inst. Washington Year Book*, **56**: 266–268.

44 Gregory, R.P.F. (1984) Circular dichroism of chlorophylls in protein complexes from chloroplasts of higher plants: a critical assessment. In Blauer, G. & Sund, H., eds, *Optical Properties and Structure of Tetrapyrroles*, de Gruyter, Berlin, pp. 475–488.

45 Murata, N., Nishimura, M. & Takamiya, A. (1966) Fluorescence of chlorophyll in photosynthetic systems. III. Emission and action spectra of fluorescence – three emission bands of chlorophyll a and the energy transfer between the two pigment systems. *Biochim. Biophys. Acta* **126**: 234–243.

46 Wittmershaus, B.P. (1987) Measurements and kinetic modelling of picosecond time-resolved fluorescence from photosystem I and chloroplasts. In Biggins, J., ed., *Progress in Photosynthesis Research*, Proc. VIIth Int. Congr. Photosynth. 1986, Nijhoff, Dordrecht, vol. 1, pp. 75–82.

47 Mahler, H. & Cordes, E.H. (1966) *Biological Chemistry*, Harper and Row, New York, p. 208.

48 Seely, G.R. (1978) The energetics of electron-transfer reactions of chlorophyll and other compounds. *Photochem. Photobiol.* **27**: 639–654.

49 Mortenson, L.E., Valentine, R.E. & Carnahan, J.E. (1962) An electron transport factor from *Clostridium pasteurianum. Biochem. Biophys. Res. Comm.* **7**: 448–452.

50 Rieske, J.S., MacLennon, D.H. & Coleman, R. (1964) Isolation and properties of an iron-protein from the (reduced coenzyme Q)-cytochrome c reductase complex of the respiratory chain. *Biochem. Biophys. Res. Comm.* **15**: 338–344.

51 Green, D.E. & Wharton, D.C. (1963) Stoichiometry of the fixed oxidation–

reduction components of the electron-transport chain of beef heart mitochondria. *Biochem. Z*. **338**: 335–348.

52 Wessels, J.S.C. (1966) Isolation of a chloroplast fragment fraction with $NADP^+$ photoreducing activity dependent on plastocyanin and independent of cytochrome f. *Biochim. Biophys. Acta* **126**: 581–583.

53 Nugent, J.H.A. & Bendall, D.S. (1987) Functional size measurements on the chloroplast bf complex. *Biochim. Biophys. Acta* **893**: 177–183.

54 Barber, J. (1984) Further evidence for the common ancestry of cytochrome b–c complexes. *Trends Biochem. Sci*. **9**: 209–211.

55 Jones, R.W. & Whitmarsh, J. (1987) The electrogenic reaction and proton release during quinol oxidation by the cytochrome b/f complex. In Biggins, J., ed., *Progress in Photosynthesis Research*, Proc. VIIth Int. Congr. Photosynth. 1986, Nijhoff, Dordrecht, vol. 2, pp. 445–452.

56 Mitchell, P. (1976) Possible molecular mechanisms of the protonmotive functions of cytochrome systems. *J. Theoret. Biol*. **62**: 327–367.

57 Moss, D.A. & Bendall, D.S. (1984) Cyclic electron transport in chloroplasts. The Q-cycle and the site of action of antimycin A. *Biochim. Biophys. Acta* **767**: 389–395.

58 Garab, G. and Hind, G. (1987) Cyclic electron transport around photosystem I in washed thylakoids. In Biggins, J., ed., *Progress in Photosynthesis Research*, Proc. VIIth Int. Congr. Photosynth. 1986, Nijhoff, Dordrecht, vol. 2, pp. 541–544.

59 Lowe, A.G. & Jones, M.N. (1984) Proton motive force—what price Δp? *Trends Biochem. Sci*. **9**: 11–12.

60 Rumberg, B. & Siggel, U. (1969) pH changes in the inner phase of the thylakoids during photosynthesis. *Naturwissenschaften* **56**: 130–132.

61 Tran-Anh, T. & Rumberg, B. (1987) Coupling mechanism between proton transport and ATP synthesis in chloroplasts. In Biggins, J., ed., *Progress in Photosynthesis Research*. Proc. VIIth Int. Congr. Photosynth. 1986, Nijhoff, Dordrecht, vol. 3, pp. 185–188.

62 Stoeckenius, W. (1985) The rhodopsin-like pigments of halobacteria: light-energy and signal transducers in an archaebacterium. *Trends Biochem. Sci*. **10**: 483–486.

63 Westerhoff, H.V. & Dancshazy, Zs. (1984) Keeping a light-driven proton-pump under control. *Trends Biochem. Sci*. **9**: 112–117.

64 Stoeckenius, W. & Bogolmoni, R.A. (1982) Bacteriorhodopsin and related pigments of halobacteria. *Annu. Rev. Biochem*. **52**: 587–615.

65 Hader, D-P. & Tevini, M. (1987) *General Photobiology*. Pergamon, Oxford, Chap. 8.

66 Dutton, P.L. (1986) Energy transduction in anoxygenic photosynthesis. In Staehelin, L.A. & Arntzen, C.J., eds, Photosynthesis III. *Encyclopaedia of Plant Physiol*., new series, vol. 19, Springer-Verlag, Berlin, pp. 197–237.

67 Cogdell, R.J. (1986) Light-harvesting complexes in the purple bacteria. In Staehelin, L.A. & Arntzen, C.J., eds, Photosynthesis III, *Encyclopaedia of Plant Physiol*., new series, vol. 19, Springer-Verlag, Berlin, pp. 252–259.

68 Michel, H. & Deisenhofer, J. (1986) X-ray diffraction studies on a crystalline bacterial photosynthetic reaction center: a progress report and conclusions on the structure of photosystem II reaction centres. In Staehelin, L.A. & Arntzen, C.J., eds, Photosynthesis III. *Encyclopaedia of Plant Physiol*., new series, vol. 19, Springer-Verlag, Berlin, pp. 371–381.

69 Thornber, J.P. (1970) Photochemical reactions of purple bacteria as revealed by studies of three spectrally different carotenobacteriochlorophyll–protein complexes isolated from *Chromatium*, strain D. *Biochemistry* **9**: 2688-2698.

70 Wasiliewsky, M.R. & Tiede, D.M. (1986) Sub-picosecond measurements of primary electron transfer in *Rhodopseudomonas viridis* reaction centres using near-infrared excitation. *FEBS Lett.* **204**: 368–372.

71 Norris, J.R. & van Brakel, G. (1986) Energy trapping in photosynthesis as probed by the magnetic properties of reaction centers. In Staehelin, L.A. & Arntzen, C.J., eds, Photosynthesis III. *Encyclopaedia of Plant Physiol.*, new series, vol. 19, Springer-Verlag, Berlin, pp. 353–370.

72 Parson, W.W. & Holten, D. (1986) Primary electron transfer reactions in photosynthetic bacteria: energetics and kinetics of transient states. In Staehelin, L.A. & Arntzen, C.J., eds, Photosynthesis III. *Encyclopaedia of Plant Physiol.*, new series, vol. 19, Springer-Verlag, Berlin, pp. 338–343.

73 Deisenhofer, J., Epp, O., Miki, K., Huber, R. & Michel, H. (1985) Structure of the protein subunits in the photosynthetic reaction center of *Rhodopseudomonas viridis* at 3A resolution. *Nature* **318**: 618–624.

74 Jackson, J.B. & Dutton, P.L. (1973) The kinetics and redox potentiometric resolution of the carotenoid shifts in *Rhodopseudomonas sphaeroides* chromatophores: their relationship to electric fields alterations in electron transport and energy coupling. *Biochim. Biophys. Acta* **325**: 102–113.

75 Petty, K.M. & Jackson, J.B. (1979) Two protons are transferred per ATP synthesized after flash activation of chromatophores from photosynthetic bacteria. *FEBS Lett.* **97**: 367–372.

76 Petty, K.M. & Jackson, J.B. (1979) Correlation between ATP synthesis and the decay of the carotenoid bandshift after single turnover flash activation of chromatophores from *Rps. capsulata. Biochim. Biophys. Acta* **547**: 463–473.

77 Melandri, B.A. & Venturoli, G. (1986) Local and delocalized interactions in energy coupling. In Staehelin, L.A. & Arntzen, C.J., eds, Photosynthesis III. *Encyclopaedia of Plant Physiol.*, new series, vol. 19, Springer-Verlag, Berlin, pp. 560–569.

78 Frenkel, A.W. (1956) Photophosphorylation of adenine nucleotides by cell-free preparations of purple bacteria. *J. Biol. Chem.* **222**: 823–834.

79 Keister, D.L. & Yike, N.J. (1967) Energy-linked reactions in photosynthetic bacteria. I Succinate-linked ATP-driven NAD^+ reduction by *Rhodospirillum rubrum* chromatophores. *Arch. Biochem. Biophys.* **121**: 415–422.

80 Drews, G. (1986) Adaptation of the bacterial photosynthetic apparatus to different light intensities. *Trends Biochem. Sci.* **11**: 255–257.

81 Ferguson, S.J. (1987) Cytochrome c springs a surprise. *Trends Biochem. Sci.* **12**: 124–125.

82 Urbach, W. (1979) Eukaryotic algae. In Trebst, A. & Avron, M., eds, Photosynthesis I. *Encyclopaedia of Plant Physiol*, new series, vol. 5, Springer-Verlag, Berlin, pp. 605–624.

83 Emerson, R. (1958) The quantum yield of photosynthesis. *Ann. Rev. Plant Physiol.* **9**: 1–24.

84 Blinks, L.R. (1957) Chromatic transients in photosynthesis of red algae. In Gaffron, H. et al, eds, *Research in Photosynthesis*, Interscience, New York, pp. 444–449.

85 Duysens, L.N.M., Amesz, J. & Kamp, B.M. (1961) Two photochemical systems in photosynthesis. *Nature*, **190**: 510–511.

86 Kautsky, H., Appel, W. & Amman, N. (1960). Chlorophyll fluorescence and carbonic acid assimilation. *Biochem. Z.* **332**: 277–292.

87 Kok, B. & Gott, W. (1960) Activation spectrum of 700 mμ absorption change in photosynthesis. *Plant Physiol.* **35**: 802–808.

88 Hill, R. & Bendall, F. (1960) Function of the two cytochrome components in chloroplasts: a working hypothesis. *Nature* **186**: 136–137.

89 Hill, R. (1939) Oxygen produced by isolated chloroplasts. *Proc. Roy. Soc. B* **127**: 192–210.

90 Giaquinta, R.T. & Dilley, R.A. (1975) A partial reaction in photosystem II: reduction of silicomolybdate prior to the site of dichlorophenyldimethylurea inhibition. *Biochim. Biophys. Acta* **387**: 288–305.

91 Joliot, P., Barbieri, G. & Chabaud, R. (1969) Un nouveau modèle des centres photochimiques du système II. *Photochem. Photobiol.* **10**: 309–329.

92 Kok, B., Forbush, B. & McGloin, M. (1970) Cooperation of charges in photosynthetic oxygen evolution—I. A linear four-step mechanism. *Photochem. Photobiol.* **11**: 457–475.

93 Kautsky, H. & Hirsch, A. (1931) Neue Versuche zur Kohlensaureassimilation. *Naturwissenschaften* **19**: 964.

94 Strehler, B.L. & Arnold, W. (1951) Light production by green plants. *Arch. Biochem. Biophys.* **34**: 809–829.

95 Malkin, S. (1977) Delayed luminescence. In Trebst, A. & Avron, M., eds, Photosynthesis I. *Encyclopaedia of Plant Physiol*, new series, vol. 5, Springer-Verlag, Berlin, pp. 473–491.

96 Renger, G. (1972) The action of 2-anilinothiophenes as accelerators of the deactivation reactions of the watersplitting enzyme system of photosynthesis. *Biochim. Biophys. Acta* **256**: 428–439.

97 Camm, E.L. & Green, B.R. (1983) Relationship between the two minor chlorophyll a–protein complexes and the photosystem II reaction center. *Biochim. Biophys. Acta* **724**: 291–293.

98 Diner, B.A. & Wollman, F-A. (1980) Isolation of highly active photosystem II particles from a mutant of *Chlamydomonas reinhardtii. Eur. J. Biochem.* **110**: 521–526.

99 Kyle, D.J. & Ohad, I. (1986) The mechanism of photoinhibition in higher plants and green algae. In Staehelin, L.A. & Arntzen, C.J., eds, Photosynthesis III. *Encyclopaedia of Plant Physiol.*, new series, vol. 19, Springer-Verlag, Berlin, pp. 468–475.

100 Bendall, D.S. (1968) Oxidation–reduction potentials of cytochromes in chloroplasts from higher plants. *Biochem. J.* **109**: 46P–47P.

101 Bendall, D.S. (1982) Photosynthetic cytochromes of oxygenic organisms. *Biochim. Biophys. Acta* **683**: 119–151.

102 Bergström, J. & Franzen, L.G. (1987) Restoration of high-potential cytochrome b-559 in salt-washed photosystem II-enriched membranes as revealed by EPR. *Acta Chem. Scand.* series B, **B41**: 126–128.

103 Döring, G., Renger, G., Vater, J. & Witt, H.T. (1969) Properties of the photoactive chlorophyll-aII in photosynthesis. *Z. Naturforsch.* **24b**: 1139–1143.

104 Hoff, A.J. (1986) Optically detected magnetic resonance (ODMR) of triplet states in vivo. In Staehelin, L.A. & Arntzen, C.J., eds, Photosynthesis III. *Encyclopaedia of Plant Physiol*, new series, vol. 19, Springer-Verlag, Berlin, pp. 400–421.

105 Klimov, V.V. & Krasnowskii, A.A. (1981) Phaeophytin as the primary electron acceptor in photosystem 2 reaction centres. *Photosynthetica* **15**: 592–609.

106 Ganago, I.B., Klimov, V.V., Ganago, A.O., Shuvalov, V.A. & Erokhin, Y.E. (1982) Linear dichroism and orientation of phaeophytin, the intermediary acceptor in photosystem II reaction centers. *FEBS Lett.* **140**: 127–130.

107 Klimov, V.V., Allakhverdiev, S.I. & Paschenko, V.Z. (1978) Measurement of the activation energy and lifetime of the fluorescence of photosystem 2 chlorophyll. *Dokl. Akad. Nauk. USSR* **242**: 1204–1207.

108 Shuvalov, V.A. & Klimov, V.V. (1976) The primary photoreactions in the complex

cytochrome-P890-P760 (bacteriophaeophytin-760) of *Chromatium minutissimum* at low redox potentials. *Biochim. Biophys. Acta* **440**: 587–599.

109 Moya, I., Hodges, M. & Briantais, J-M. (1987) Is variable fluorescence due to charge recombination? In Biggins, J., ed., *Progress in Photosynthesis Research*, Proc. VIIth Int. Congr. Photosynth. 1986, Nijhoff, Dordrecht, vol. 1, pp. 111–114.

110 Duysens, L.N.M. & Sweers, H.E. (1963) Mechanism of two photochemical reactions in algae as studied by fluorescence. In Miyachi, S., ed., *Studies on Microalgae and Photosynthetic Bacteria*, Univ. Tokyo Press, Tokyo, pp. 353–372.

111 Stiehl, H.H. & Witt, H.T. (1968) Die kurtzzeitigen ultravioletten Differenzspektren bei der Photosynthese. *Z. Naturforsch.* **23b**: 220–224.

112 Bensasson, R. & Land, E.J. (1973) Optical and kinetic properties of semireduced plastoquinone and ubiquinone: electron acceptors in photosynthesis. *Biochim. Biophys. Acta* **325**: 175–181.

113 van Gorkom, H.J. (1974) Identification of the reduced primary electron acceptor of photosystem II as a bound semiquinone. *Biochim. Biophys. Acta* **347**: 439–442.

114 Knaff, D.B. & Arnon, D.I. (1969) Spectral evidence for a new photoactive component of the oxygen-evolving system in photosynthesis. *Proc. Natl Acad. Sci. USA* **63**: 963–969.

115 Cox, R.P. & Bendall, D.S. (1974) The functions of plastoquinone and β-carotene in photosystem II of chloroplasts. *Biochim. Biophys. Acta* **347**: 49–59.

116 Klimov, V.V., Dolan, E., Shaw, E.R. & Ke, B. (1980) Interaction between the intermediary electron acceptor (phaeophytin) and a possible plastoquinone–iron complex in photosystem II reaction centers. *Proc. Natl Acad. Sci. USA* **77**: 7227–7231.

117 Ben-Hayyim, G. & Neumann, J. (1975) On the mechanism of action of silico-molybdic acid in chloroplasts. *FEBS Lett.* **56**: 240–243.

118 Govindjee & Eaton-Rye, J.J. (1986) Electron transfer through photosystem II acceptors: interaction with anions. *Photosynthesis Res.* **10**: 365–379.

119 Diner, B.A. (1986) The reaction center of photosystem II. In Staehelin, L.A. & Arntzen, C.J., eds, Photosynthesis III. *Encyclopaedia of Plant Physiol.*, new series, vol. 19, Springer-Verlag, Berlin, pp. 422–436.

120 O'Malley, P.J., Babcock, G.T. & Prince, R.T. (1984) The cationic plastoquinone radical of the chloroplast water splitting complex. Hyperfine splitting from a single methyl group determines the spectral shape of signal II. *Biochim. Biophys. Acta* **766**: 283–288.

121 Babcock, G.T., Chandrashekar, T.J., Ghanotakis, D.F., Hoganson, D.W., O'Malley, P.J., Rodruigez, I.D. & Yocum, C.F. (1987) Kinetics and structure on the high potential side of photosystem II. In Biggins, J., ed., *Progress in Photosynthesis Research*, Proc. VIIth Int. Congr. Photosynth. 1986, Nijhoff, Dordrecht, vol. 1, pp. 463–469.

122 Yocum, C.F., Yerkes, C.T., Blankenship, R.E., Sharp, R.R. & Babcock, G.T. (1981) Stoichiometry, inhibitor sensitivity, and organization of manganese associated with photosynthetic oxygen evolution. *Proc. Natl Acad. Sci. USA* **78**: 7507–7511.

123 Yocum, C.F. (1986) Electron transfer on the oxidising side of photosystem II: components and mechanism. In Staehelin, L.A. & Arntzen, C.J., eds, Photosynthesis III. *Encyclopaedia of Plant Physiol.*, new series, vol. 19, Springer-Verlag, Berlin, pp. 437–446.

124 Åkerlund, H.-E. (1981) Partial reconstitution of the photosynthetic water splitting in inside out thylakoid vesicles. In Akoyunoglou, G., ed., *Photosynthesis*. Proc. Vth Int. Congr. Photosynth. 1980, Balaban, Philadelphia, vol. 2, pp. 465–472.

125 Brudvig, G.W. & de Paula, J.C. (1987) On the mechanism of photosynthetic water oxidation. In Biggins, J., ed., *Progress in Photosynthesis Research*, Proc. VIIth Int. Congr. Photosynth. 1986, Nijhoff, Dordrecht, vol, 1, pp. 491–498.

126 Coleman, W.J. & Govindjee (1987) A model for the mechanism of chloride activation of oxygen evolution in photosystem II. *Photosynthesis Res.* **13**: 199–223.

127 Melis, A. & Homann, P.H. (1976) Heterogeneity of the photochemical centers in system II of chloroplasts. *Photochem. Photobiol.* **23**: 343–350.

128 Melis, A. & Homann, P.H. (1978) A selective effect of Mg^{2+} on the photochemistry at one type of reaction center in photosystem II of chloroplasts. *Arch. Biochem. Biophys.* **190**: 523–530.

129 Boardman, N.K. & Anderson, J.M. (1964) Isolation from spinach chloroplasts of particles containing different proportions of chlorophyll a and b and their possible role in photosynthesis. *Nature* **293**: 166–167.

130 Stewart, A.C. & Bendall, D.S. (1981) Properties of oxygen-evolving photosystem-II particles from *Phormidium laminosum*, a thermophilic blue-green alga. *Biochem. J.* **194**: 877–887.

131 Schatz, G.H. & Witt, H.T. (1984) Extraction and characterization of oxygen-evolving photosystem II complexes from a thermophilic cyanobacterium *Synechococcus* spec. *Photobiochem. Photobiophys.* **7**: 1–14.

132 Rogner, M., Dekker, J.P., Boekema, E.J. & Witt, H.T. (1987) Size, shape and mass of the oxygen-evolving photosystem II complex from the thermophilic cyanobacterium *Synechococcus* sp. *FEBS Lett.* **219**: 207–211.

133 Berthold, D.A., Babcock, G.T. & Yocum, C.F. (1981) A highly resolved, oxygen-evolving photosystem II preparation from spinach thylakoid membranes. *FEBS Lett.* **134**: 231–234.

134 Franzen, L.G. (1987) Isolation and characterization of an oxygen-evolving photosystem II reaction center complex from spinach. *Acta Chem. Scand.* ser. B, **41**: 103–105.

135 Camm, E.L. & Green, B.R. (1980) Fractionation of thylakoid membranes with the non-ionic detergent octyl-D-glucopyranoside. *Plant Physiol.* **66**: 428–432.

136 Picaud, A., Acker, S. & Duranton, J. (1982) A single-step separation of PS1, PS2 and chlorophyll-antenna particles from spinach chloroplasts. *Photosynthesis Res.* **3**: 202–213.

137 Kok, B. (1956) On the reversible absorption change at 705 mμ in photosynthetic organisms. *Biochim. Biophys. Acta* **22**: 399–401.

138 Arnon, D.I., Allen, M.B. & Whatley, F.R. (1954) Photosynthesis by isolated chloroplasts. *Nature* **174**: 394–396.

139 Arnon, D.I., Whatley, F.R. & Allen, M.B. (1958) Assimilatory power in photosynthesis. *Science* **127**: 1026–1034.

140 Trebst, A. (1974) Energy conservation in photosynthetic electron transport of chloroplasts. *Annu. Rev. Plant Physiol.* **25**: 423–458.

141 West, K.R. & Wiskich, J.T. (1968) Photosynthetic control by isolated pea chloroplasts. *Biochem. J.* **109**: 527–532.

142 Heber, U. & Kirk, M.R. (1974) Flexibility of coupling and stoichiometry of ATP formation in intact chloroplasts. *Biochim. Biophys. Acta* **376**: 136–150.

143 Wessels, J.S.C. (1977) Fragmentation. In Trebst, A. & Avron, M., eds, Photosynthesis I. *Encyclopaedia of Plant Physiol.*, new series, vol. 5, Springer-Verlag, Berlin, pp. 563–573.

144 Vernon, L.P. & Shaw, E.R. (1971) Subchloroplast fragments: Triton X-100 method. *Methods Enzymol.* **23A**: 277–289.

145 Thornber, J.P., Gregory, R.P.F., Smith, C.A. & Bailey, J.L. (1967) Studies on

the nature of the chloroplast lamella. I. Preparation and some properties of two chlorophyll–protein complexes. *Biochemistry* **6**: 391–396.

146 Gray, J.C., Blyden, E.R., Eccles, C.J., Dunn, P.P.J., Hird, S.M., Hoglund, A-S., Kaethner, T.M., Smith, A.G., Willey, D.L. & Dyer, T.A. (1987) Chloroplast genes for photosynthetic membrane components. In Biggins, J., ed., *Progress in Photosynthesis Research*. Proc. VIIth Int. Congr. Photosynth. 1986, Nijhoff, Dordrecht, vol. 4, pp. 617–624.

147 Boekema, E.J., Dekker, J.P., van Heel, M.G., Rogner, M., Saenger, W., Witt, I. & Witt, H.T. (1987) Evidence for a trimeric organization of the photosystem I complex from the thermophilic cyanobacterium *Synechococcus* sp. *FEBS Lett.* **217**: 283–286.

148 Ikegami, I. & Ke, B. (1984) A 160-kilodalton photosystem I reaction-center complex. Low-temperature absorption and EPR spectroscopy of the early electron acceptors. *Biochim. Biophys. Acta* **764**: 70–79.

149 Alberte, R.S. & Thornber, J.P. (1978) Rapid procedure for isolating photosystem I reaction centers in a highly enriched form. *FEBS Lett.* **91**: 126–130.

150 Sétif, P. & Mathis, P. (1980) The oxidation–reduction potential of P700 in chloroplast lamellae and subchloroplast particles. *Arch. Biochem. Biophys.* **204**: 477–485.

151 Kok, B. (1961) Partial purification and determination of oxidation–reduction potential of the photosynthetic chlorophyll complex absorbing at 700 mμ. *Biochim. Biophys. Acta* **48**: 527–533.

152 Philipson, K.D., Sato, V.L. & Sauer, K. (1972) Exciton interaction in the photosystem I reaction center from spinach chloroplasts. Absorption and circular dichroism difference spectra. *Biochemistry* **11**: 4591–4595.

153 Dörnemann, D. & Senger, H. (1982) Physical and chemical properties of chlorophyll RCI extracted from photosystem I of spinach leaves and from algae. *Photochem. Photobiol.* **35**: 821–826.

154 Watanabe, T., Kobayashi, M., Nakazato, M., Ikegami, I. & Hiyama, T. (1987) Chlorophyll a′ in photosynthetic apparatus: reinvestigation. In Biggins, J., ed., *Progress in Photosynthesis Research*. Proc. VIIth Int. Congr. Photosynth. 1986, Nijhoff, Dordrecht, vol. 1, pp. 303–306.

155 Evans, M.C.W. (1982) Iron sulfur centres in photosynthetic electron transport. *Met. Ions Biol.* **4**: 249–284.

156 Oh-Oka, H., Takahashi, Y., Wada, K., Matsubara, H., Ohyama, K. & Ozeki, H. (1987) The 8 kDa polypeptide in photosystem I is a probable candidate of an iron-sulfur center protein coded by the chloroplast gene frxA. *FEBS Lett.* **218**: 52–54.

157 Kukarskikh, G.P., Plutakhin, G.A., Tulbu, G.A. & Krendeleva, T.E. (1987) Pathways of photoinduced electron transport on the acceptor side of photosystem I. *Biokhimiya (Moscow)* **52**: 927–933, cited *Chem. Abst.* **107**: 93783.

158 Sauer, K., Mathis, P., Acker, S. & van Best, J.A. (1978) Electron acceptors associated with P-700 in Triton solubilized photosystem I particles from spinach chloroplasts. *Biochim. Biophys. Acta* **503**: 120–134.

159 Sétif, P., Mathis, P., Lagoutte, B. & Duranton, J. (1983) Electron acceptors in PS-I: comparison of A2 and X. In Sybesma, C., ed., *Advances in Photosynthesis Research*, Proc. VIth Int. Congr. Photosynth. 1983, Nijhoff, The Hague, vol. 2, pp. 589–591.

160 Shuvalov, V.A., Dolan, E. & Ke, B. (1979) Spectral and kinetic evidence for two early acceptors in photosystem I. *Proc. Natl Acad. Sci. USA* **76**: 770–773.

161 Ziegler, K., Lockau, W. & Nitschke, W. (1987) Bound electron acceptors of pho-

tosystem I. Evidence against the identity of redox center A1 with phylloquinone. *FEBS Lett*. **217**: 16–20.

162 Katoh, S. (1960) A new copper protein from *Chlorella ellipsoidea*. *Nature* **186**: 533–534.

163 Guss, J.M. & Freeman, H.C. (1983) Structure of oxidized poplar plastocyanin at 1.6 A resolution. *J. Mol. Biol*. **169**: 521–563.

164 Haehnel, W. (1986) Plastocyanin. In Staehelin, L.A. & Arntzen, C.J., eds, Photosynthesis III. *Encyclopaedia of Plant Physiol*, new series, vol. 19, Springer-Verlag, Berlin, pp. 547–559.

165 Morrissey, P.J., McCauley, S.W. & Melis, A. (1987) Differential solubilization of the integral electron transport complexes from the thylakoid membrane of spinach chloroplasts. Localization of photosystem I, photosystem II and the cytochrome b6-f complex. In Biggins, J., ed., *Progress in Photosynthesis Research*. Proc. VIIth Int. Congr. Photosynth. 1986, Nijhoff, Dordrecht, vol. 2, 305–308.

166 Whitmarsh, J. (1986) Mobile electron carriers in thylakoids. In Staehelin, L.A. & Arntzen, C.J., eds, Photosynthesis III. *Encyclopaedia of Plant Physiol*., new series, vol. 19, Springer-Verlag, Berlin, pp. 508–527.

167 Hauska, G. (1986) Composition and structure of cytochrome bc1 and b6f complexes. In Staehelin, L.A. & Arntzen, C.J., eds, Photosynthesis III. *Encyclopaedia of Plant Physiol*., new series, vol. 19, Springer-Verlag, Berlin, pp. 496–507.

168 Pierson, B.K. & Castenholz, R.W. (1974) Phototrophic gliding filamentous bacterium of hot springs, *Chloroflexus aurantiacus*. *Arch. Microbiol*. **100**: 5–24, 283–305.

169 Gibson, J., Ludwig, W., Stackebrandt, E. & Woese, C.R. (1985) The phylogeny of the green photosynthetic bacteria: absence of a close relationship between *Chlorobium* and *Chloroflexus*. *Syst. Appl. Microbiol*. **6**: 152–156.

170 Olson, J.M., Prince, R.C. & Brune, D.C. (1977) Reaction-center complexes from green bacteria. In Olson, J.M. and Hind, G., eds, *Chlorophyll-Proteins, Reaction Centers and Photosynthetic Membranes*, Brookhaven Symp. Biol., 28, pp. 238–246.

171 Amesz, J. (1984) Electron transport in green bacteria. In Sybesma, C., ed., *Advances in Photosynthesis Research*, Proc. VIth Int. Congr. Photosynth. 1983, Nijhoff, The Hague, vol. 1, pp. 621–628.

172 Nitschke, W., Feiler, U., Lockau, W. & Hauska, G. (1987) The photosystem of the green sulphur bacterium *Chlorobium limicola* contains two early electron acceptors similar to photosystem I. *FEBS Lett*. **218**: 283–286.

173 Blankenship, R.E. & Fuller, R.C. (1986) Membrane topology and photochemistry of the green photosynthetic bacterium *Chloroflexus aurantiacus*. In Staehelin, L.A. & Arntzen, C.J., eds, Photosynthesis III. *Encyclopaedia of Plant Physiol*., new series, vol. 19, Springer-Verlag, Berlin, pp. 390–399.

174 Nuijs, A.M., van Dorrsen, R., Duysens, L.M.N. & Amesz, J. (1985) Excited states and primary photochemical reactions in the photosynthetic bacterium *Heliobacterium chlorum*. *Proc. Natl Acad. Sci. USA* **82**: 6965–6968.

175 Bassham, J.A., Benson, A.A., Kay, L.D., Harris, A.Z., Wilson, A.T. & Calvin, M. (1954) The path of carbon in photosynthesis. XXI. The cyclic regeneration of the carbon dioxide acceptor. *J. Amer. Chem. Soc*. **76**: 1760–1770.

176 Bassham, J.A. (1979) The reductive pentose phosphate cycle and its regulation. In Trebst, A. & Avron, M., eds, Photosynthesis I. *Encyclopaedia of Plant Physiol*, new series, vol. 5, Springer-Verlag, Berlin, pp. 9–30.

177 Consden, R., Gordon, A.H. & Martin, A.J.P. (1944) Qualitative analysis of protein: a partition chromatographic method using paper. *Biochem. J*. **38**: 224–232.

178 Jensen, R.G. & Bassham, J.A. (1966) Photosynthesis by isolated chloroplasts. *Proc. Natl Acad. Sci. USA* **56**: 1095–1101.

179 Akazawa, T. (1984) Molecular evolution of ribulose-1,5-bisphosphate carboxy-lase/oxygenase (RuBisCO). *Trends Biochem. Sci.* **9**: 380–383.

180 Salvucci, M.E., Portis, Jr, A.R. & Ogren, W.L. (1985) A soluble chloroplast protein catalyses ribulosebisphosphate carboxylase/oxygenase activation in vivo. *Photosynthesis Res.* **7**: 193–201.

181 Gutteridge, S., Parry, M.A.J., Keys, A.J., Servaites, J. & Feeney, J. (1987) The structure of the naturally occurring inhibitor of rubisco that accumulates in the chloroplast in the dark is 2′ carboxyarabinitol-1-phosphate. In Biggins, J., ed., *Progress in Photosynthesis Research*. Proc. VIIth Int. Congr. Photosynth. 1986, Nijhoff, Dordrecht, vol. 3, pp. 395–398.

182 Sue, J.M. & Knowles, J.R. (1978) Retention of the oxygens at C-2 and C-3 of D-ribulose-1,5-bisphosphate in the reaction catalysed by ribulose-1,5-bisphosphate carboxylase. *Biochemistry* **17**: 4041–4044.

183 Gibbs, M. (1963) An evaluation of the carbon reduction pathways of photosynthesis. In *Photosynthetic Mechanisms of Green Plants*. Publ. no. 1145, Natl Acad. Sci.—Natl Res. Co. Washington, D.C., USA, pp. 663–674.

184 Heber, U., Neimanis, S., Dietz, K.J. & Viil, J. (1987) Assimilatory force in relation to photosynthetic fluxes. In Biggins, J., ed., *Progress in Photosynthesis Research*, Proc. VIIth Int. Congr. Photosynth. 1986, Nijhoff, Dordrecht, vol.3, pp. 293–299.

185 Bradbeer, J.W., Ruffer-Turner, M.E., de Ferrer, E.J., Wara-Aswapati, O. & Kemble, R.J. (1981) The regulation of phosphoribulokinase and glyceraldehyde phosphate dehydrogenase ($NADP^+$) in C3 and C4 plants. In Akoyunoglou, G., ed., *Photosynthesis*, Proc. Vth Int. Congr. Photosynth. 1980, Balaban, Philadelphia, vol. 3, pp. 389–394.

186 Walker, D.A. (1981) Photosynthetic induction. In Akoyunoglou, G., ed., *Photosynthesis*. Proc. Vth Int. Congr. Photosynth. 1980, Balaban, Philadelphia, vol. 3, pp. 189–202.

187 Droux, M., Jacquot, J.-P., Miginiac-Maslow, M., Gadal, P., Crawford, N.A., Yee, B.C. & Buchanan, B.B. (1987) Ferredoxin-thioredoxin reductase: an iron-sulfur enzyme linking light to enzyme regulation in chloroplasts. In Biggins, J., ed., *Progress in Photosynthesis Research*, Proc. VIIth Int. Congr. Photosynth. 1986, Nijhoff, Dordrecht, vol. 3, pp. 249–252.

188 Preiss, J. & Levi, C. (1979) Metabolism of starch in leaves. In Gibbs, M. & Latzko, E., eds, Photosynthesis II. *Encyclopaedia of Plant Physiol.*, new series, vol. 6, Springer-Verlag, Berlin, pp. 282–312.

189 Sirevag, R., Buchanan, B.B., Berry, J.A. & Troughton, J.H. (1977) Mechanism of carbon dioxide fixation in bacterial photosynthesis studied by the carbon isotope fractionation technique. *Arch. Microbiol.* **112**: 35–38.

190 Evans, M.C.W., Buchanan, B.B. & Arnon, D.I. (1966) A new ferredoxin-dependent carbon reduction cycle in a photosynthetic bacterium. *Proc. Natl Acad. Sci. USA* **55**: 928–934.

191 Decker, J.P. (1955) A rapid post-illumination deceleration of respiration in green leaves. *Plant Physiol.* **30**: 82–84.

192 Zelitch, I. (1979) Photorespiration: studies with whole tissues. In Gibbs, M. & Latzko, E., eds, Photosynthesis II. *Encyclopaedia of Plant Physiol.*, new series, vol. 6, Springer-Verlag, Berlin, pp. 353–367.

193 Voskresenskaya, N.P. (1979) Effect of light quality on carbon metabolism. In Gibbs, M. & Latzko, E., eds, Photosynthesis II. *Encyclopaedia of Plant Physiol.*, new series, vol. 6, Springer-Verlag, Berlin, pp. 174–180.

194 Bowes, G., Ogren, W.L. & Hageman, R.H. (1971) Phosphoglycollate production catalysed by ribulose diphosphate carboxylase. *Biochem. Biophys. Res. Comm.* **45**: 716–722.

195 Tolbert, N.E. (1981) Oxidative photosynthetic carbon cycle of photorespiration. In Akoyunoglou, G., ed., *Photosynthesis.*, Proc. Vth Int. Congr. Photosynth. 1980, Balaban, Philadelphia, vol. 4, pp. 435–448.

196 Howitz, K.T. & McCarty, R.E. (1987) Tentative identification of the pea chloroplast envelope glycolate transporter. In Biggins, J., ed., *Progress in Photosynthesis Research*, Proc. VIIth Int. Congr. Photosynth. 1986, Nijhoff, Dordrecht, vol. 3, pp. 593–596.

197 Zelitch, I. (1981) Regulation of photorespiration. In Akoyunoglou, G., ed., *Photosynthesis*. Proc. Vth Int. Congr. Photosynth. 1980, Balaban, Philadelphia, vol. 4, pp. 449–461.

198 Kendall, A.C., Bright, S.W.J., Hall, N.P., Keys, A.J., Lea, P.J., Turner, J.C. & Wallsgrove, R.M. (1987) Barley photorespiration mutants. In Biggins, J., ed., *Progress in Photosynthesis Research*, Proc. VIIth Int. Congr. Photosynth. 1986, Nijhoff, Dordrecht, vol. 3, pp. 629–632.

199 Canvin, D.T. (1979) Photorespiration: comparison between C3 and C4 plants. In Gibbs, M. & Latzko, E., eds, Photosynthesis II. *Encyclopaedia of Plant Physiol*, new series, vol. 6, Springer-Verlag, Berlin, pp. 368–396.

200 Kluge, M. (1979) The flow of carbon in Crassulacean acid metabolism (CAM). In Gibbs, M. & Latzko, E., eds, Photosynthesis II. *Encyclopaedia of Plant Physiol*, new series, vol. 6, Springer-Verlag, Berlin, pp. 113–125.

201 Kortschak, H.P., Hartt, C.E. & Burr, G.O. (1965) Carbon dioxide fixation in sugarcane leaves. *Plant Physiol.* **40**: 209–213.

202 Hatch, M.D. & Slack, C.R. (1966) Photosynthesis by sugarcane leaves. A new carboxylation reaction and the pathway of sugar formation. *Biochem. J.* **101**: 103–111.

203 Troughton, J.H. (1979) $\delta13C$ as an indicator of carboxylation reactions. In Gibbs, M. & Latzko, E., eds, Photosynthesis II. *Encyclopaedia of Plant Physiol*, new series, vol. 6, Springer-Verlag, Berlin, pp. 140–149.

204 Burnell, J.N. & Hatch, M.D. (1985) Light-dark modulation of leaf pyruvate, Pi dikinase. *Trends Biochem. Sci.* **10**: 288–291.

205 Black, C.C & Mollenhauer, H.H. (1971) Structure and distribution of chloroplasts and other organelles in leaves with various rates of photosynthesis. *Plant Physiol.* **41**: 15–23.

206 Edwards, G.E. & Huber, S.C. (1979) C4 metabolism in isolated cells and protoplasts. In Gibbs, M. & Latzko, E., eds, Photosynthesis II. *Encyclopaedia of Plant Physiol*, new series, vol. 6, Springer-Verlag, Berlin, pp. 102–112.

207 Queiroz, O. (1979) Rhythms of enzyme capacity and activity as adaptive mechanisms. In Gibbs, M. & Latzko, E., eds, Photosynthesis II. *Encyclopaedia of Plant Physiol*, new series, vol. 6, Springer-Verlag, Berlin, pp. 126–139.

208 Ray, T.B. & Black, C.C. (1979) The C4 and Crassulacean acid metabolism pathways. In Gibbs, M. & Latzko, E., eds, Photosynthesis II. *Encyclopaedia of Plant Physiol*, new series, vol. 6, Springer-Verlag, Berlin, pp. 77–101.

209 Douce, R. & Joyard, J. (1979) Structure and function of the plastid envelope. *Adv. Bot. Res.* **7**: 1–116.

210 Cline, K., Andrews, J., Mersey, B., Newcomb, E.H. & Keegstra, K. (1981) Separation and characterization of inner and outer envelope membranes of pea chloroplasts. *Proc. Natl Acad. Sci. USA* **78**: 3595–3599.

211 Heldt, H.W. & Sauer, F. (1971) The inner membrane of the chloroplast envelope

as the site of specific metabolite transport. *Biochim. Biophys. Acta* **234**: 83–91.

212 Woo, K.C., Fluegge, U.I. & Heldt, H.W. (1987) A two-translocator model for the transport of 2-oxoglutarate and glutamate in chloroplasts during ammonia assimilation in the light. *Plant Physiol.* **84**: 624–632.

213 Husic, D.H., Moroney, J.V. & Tolbert, N.E. (1987) The role of carbonic anhydrase in the inorganic carbon concentrating system of *Chlamydomonas reinhardtii*. In Biggins, J., ed, *Progress in Photosynthesis Research*, Proc. VIIth Int. Congr. Photosynth. 1986, Nijhoff, Dordrecht, vol. 4, pp. 317–324.

214 Kelly, G.J. (1984) The capture of carbon by aquatic plants. *Trends Biochem. Sci.* **9**: 255–256.

215 Warburg, O., Krippahl, G. & Lehmann, A. (1969) Chlorophyll catalysis and Einstein's photochemical law in photosynthesis. *FEBS Lett.* **3**: 221–222.

216 Ellis, R.J., Highfield, P.E. & Silverthorne, J. (1978) The synthesis of chloroplast proteins by subcellular systems. In Hall, D.O., Coombs, J. & Goodwin, T.W., eds, *Photosynthesis 77*, Proc. IVth Int. Congr. Photosynth., 1977, pp. 497–506.

217 Ellis, R.J. (1981) Chloroplast proteins: synthesis, transport and assembly. *Annu. Rev. Plant Physiol.* **32**: 111–137.

218 Ellis, R.J. (1985) Synthesis, processing and assembly of polypeptide subunits of ribulose-1, 5-carboxylase/oxygenase. In Steinback, K.E., Bonitz, S., Arntzen, C.J. & Bogorad, L., eds, *Molecular Biology of the Photosynthetic Apparatus*, Cold Spring Harbor Laboratory, New York, pp. 339–347.

219 Pain, D., Kanwar, Y.S. & Blobel, G. (1988) Identification of a receptor for protein import into chloroplasts and its localization to envelope contact zones. *Nature* **331**: 232–237.

220 Stitt, M. (1987) Limitation of photosynthesis by sucrose synthesis. In Biggins, J., ed., *Progress in Photosynthesis Research*, Proc. VIIth Int. Congr. Photosynth. 1986, Nijhoff, Dordrecht, vol. 3, pp. 685–692.

221 Heldt, H.W. & Stitt, M. (1987) The regulation of sucrose synthesis in leaves. In Biggins, J., ed., *Progress in Photosynthesis Research*, Proc. VIIth Int. Congr. Photosynth. 1986, Nijhoff, Dordrecht, vol. 3, pp. 675–684.

222 Robinson, J.M. & van Berkum, P. (1987) CO_2 photofixation and NO_2-photoreduction in leaf mesophyll cell isolates from N_2-fixing soybean plants held in the absence of $NO_3 -$. In Biggins, J., ed., *Progress in Photosynthesis Research*, Proc. VIIth Int. Congr. Photosynth. 1986, Nijhoff, Dordrecht, vol. 3, pp. 545–548.

223 Kok, B. (1949) The interrelation of respiration and photosynthesis in green plants. *Biochim. Biophys. Acta* **3**: 625–631.

224 Hoch, G. & Owens, O. v. H. (1963) Photoreactions and respiration. In *Photosynthetic Mechanisms of Green Plants*. Publ. no. 1145, Natl Acad. Sci. –Natl Res. Co. Washington, D.C., pp. 409–420.

225 Wildman, S.G. & Jope, C.A. (1977) Origin of chloroplasts the size of single grana in apical meristems. *Photosynthetic Organelles, Plant & Cell Physiol.* **3**: 385–401.

226 Schon, A., Krupp, G., Gough, S., Berry-Lowe, S., Kannangara, C.G. & Soll, D. (1986) The RNA required in the first step of chlorophyll biosynthesis is a chloroplast glutamate tRNA. *Nature* **322**: 281–284.

227 Rüdiger, W., Benz, J. & Guthoff, C. (1980) Determination and partial characterization of activity of chlorophyll synthetase in etioplast membranes. *Eur. J. Biochem.* **109**: 193–200.

228 Gounaris, K. & Barber, J. (1983) Monogalactosyldiacyglycerol: the most abundant polar lipid in Nature. *Trends Biochem. Sci.* **8**: 378–381.

229 Roughan, G. & Slack, R. (1984) Glycerolipid synthesis in leaves. *Trends Biochem. Sci.* **9**: 383–386.

230 Frentzen, M., Heinz, E., McKeon, T.A. & Stumpf, P.K. (1983) Specificities and selectivities of glycerol-3-phosphate acyltransferase and monoacylglycerol-3-phosphate acyltransferase from pea and spinach chloroplasts. *Eur. J. Biochem.* **129**: 629–636.

231 Keegstra, K. (1986) The biosynthesis of chloroplast membrane lipids. In Staehelin, L.A. & Arntzen, C.J., eds, Photosynthesis III. *Encyclopaedia of Plant Physiol*, new series, vol. 19, Springer-Verlag, Berlin, pp. 706–712.

232 Siegenthaler, P.A. & Rawyler, A. (1986) Acyl lipids in thylakoid membranes: distribution and involvement in photosynthetic functions. In Staehelin, L.A. & Arntzen, C.J., eds, Photosynthesis III. *Encyclopaedia of Plant Physiol*, new series, vol. 19, Springer-Verlag, Berlin, pp. 693–705.

233 Radunz, A., Bader, K.P. & Schmid, G.H. (1984) Influence of antiserums to sulfoquinovosyl diglyceride and to β-sitosterol on the photosynthetic electron transport in chloroplasts from higher plants. *Dev. Plant Biol.* **9**: 479–484.

234 Hanley-Bowdoin, L. & Chua, N.-H. (1987) Chloroplast promoters. *Trends Biochem. Sci.* **12**: 67–70.

235 Ohyama, K., Fukuzawa, H., Kohchi, T., Shirai, H., Sana, T., Sano, S., Umesono, K., Shiki, Y., Takeuchi, M., Chang, Z., Aota, S., Inokuchi, H. & Ozeki, H. (1986) Chloroplast gene organisation deduced from complete sequence of liverwort *Marchantia polymorpha* chloroplast DNA. *Nature* **322**: 572–574.

236 Shinozaki, K., Ohme, M., Tanaka, M., Wakasugi, T., Hayashida, N., Matsubayishi, T., Zaita, N., Chunwongse, J., Obokata, J., Yamaguchi-Shinozaki, K., Ohto, C., Torazawa, K., Meng, B.Y., Sugita, M., Deno, H., Kamogashira, T., Yamada, K., Kusuda, J., Takaiwa, F., Kato, A., Tohdoh, N., Shimada, H. & Sugiura, M. (1986) The complete nucleotide sequence of the tobacco chloroplast genome: gene organisation and expression. *EMBO J.* **5**: 2043–2049.

237 Ohyama, K., Kohchi, T., Sano, T. & Yamada, Y. (1988) Newly identified groups of genes in chloroplasts. *Trends Biochem. Sci.* **13**: 19–23.

238 Rasmussen, O.F., Bookjans, G., Stumman, B.M. & Henningsen, K.W. (1984) Localization and nucleotide sequence of the gene for the membrane polypeptide D2 from pea chloroplast DNA. *Plant Mol. Biol.* 191–199.

239 Rochaix, J.D., Dron, M., Rahire, M. & Malnoe, P. (1984) Sequence homology between the 32K dalton and the D2 chloroplast membrane polypeptides of *Chlamydomonas reinhardtii*. *Plant Mol. Biol.* **3**: 363–370.

240 Michalski, T.J., Hunt, J.E., Bowman, M.K., Smith, U., Bardeen, K., Gest, H., Norris, J.R. & Katz, J.J. (1987) Bacteriophaeophytin g: properties and some speculations on a primary role for bacteriochlorophylls b and g in the biosynthesis of chlorophylls. *Proc. Natl Acad. Sci. USA* **84**: 2570–2574.

241 Schirmer, T., Bode, W. & Huber, R. (1987) Refined three-dimensional structures of two cyanobacterial C-phycocyanins at 2.1 and 2.5 A resolution. A common principle of phycobilin–protein interaction. *J. Mol. Biol.* **196**: 677–695.

242 Hoyer-Hansen, G., Bassi, R., Hønberg, L.S. & Simpson, D.J. (1988) Immunological characterization of chlorophyll a/b-binding proteins of barley thylakoids. *Planta* **173**: 12–21.

243 Allen, J.P., Feher, G., Yeates, T.O., Komiya, H. & Rees, D.C. (1987) Structure of the reaction centre from *Rhodobacter spheroides* R-26: the protein subunits. *Proc. Natl Acad. Sci. USA* **84**: 6162–6166.

244 Barry, B.A. & Babcock, G.T. (1987) Tyrosine radicals are involved in the photosynthetic oxygen-evolving system. *Proc. Natl Acad. Sci. USA* **84**: 7099–7103.

245 Debus, R.J., Barry, B.A., Babcock, G.T. & McIntosh, L. (1988) Site-directed mutagenesis identifies a tyrosine radical involved in the photosynthetic oxygen-evolving

system. *Proc. Natl Acad. Sci. USA* **85**: 427–430.

246 Joyard, J. & Douce, R. (1988) Import receptor in chloroplast envelope. Comments. *Nature* **333**: 306–307.

247 Vermass, W.F.J., Williams, J.G.K., Chisholm, D.A. & Arntzen, C.J. (1987) Site-directed mutagenesis in the photosystem II gene *psb*D, encoding the D2 protein. In Biggins, J., ed., *Progress in Photosynthesis Research*. Proc. VIIth Int. Congr. Photosynth. 1986, Nijhoff, Dordrecht, vol. 4, pp. 805–808.

248 Hall, D. O. (1972) Nomenclature for isolated chloroplasts. *Nature* **235**: 125–126.

249 Jensen, R.G. (1979) The isolation of intact leaf cells, protoplasts and chloroplasts. In Gibbs, M. & Latzko, E., eds, Photosynthesis II. *Encyclopaedia of Plant Physiol.*, new series, vol. 6. Springer-Verlag, Berlin. pp. 31–40.

250 Markwell, J.P., Reinman, S. & Thornber, J.P. (1978) Chlorophyll–protein complexes from higher plants—a procedure for improved stability and fractionation. *Arch. Biochem. Biophys.* **190**: 136–141.

251 Machold, O., Simpson, D J. & Lindberg-Moller, B. (1979) Chlorophyll–proteins of thylakoids from wild type and mutants of barley (*Hordeum vulgare* L.). *Carlsberg Res. Commun.* **44**: 235–254.

252 Anderson, J.M., Waldron, J.C. & Thorne, S.W. (1978) Chlorophyll–protein complexes of spinach and barley thylakoids. Spectral characteristics of six complexes resolved by an improved electrophoretic procedure. *FEBS Lett.* **92**: 227–233.

253 Delepelaire, P. & Chua, N.-H. (1981) Electrophoretic purification of chlorophyll a/b–protein complexes from *Chlamydomonas reinhardtii* and spinach and analysis of their polypeptide compositions. *J. Biol. Chem.* **256**: 9300–9307.

254 Dunahay, T.G., Staehelin, L.A., Seibert, M., Ogilvie, P.D. & Berg, S.P. (1984) Structural, biochemical and biophysical characterization of four oxygen-evolving photosystem II preparations from spinach. *Biochim. Biophys. Acta* **764**: 179–193.

255 Markwell, J.P., Thornber, J.P. & Boggs, R.T. (1979) Higher plant chloroplasts: evidence that all the chlorophyll exists as chlorophyll–protein complexes. *Proc. Natl. Acad. Sci. USA* **76**: 1233–1235.

256 Crouse, E.J., Schmitt, J.M. & Bohnert, H.-J. (1985) Chloroplast and cyanobacterial genomes: a compilation. *Plant Mol. Biol. Reporter* **3**: 43–89.

257 Godde, D. (1982) Evidence for a membrane bound NADH-plastoquinone-oxidoreductase in *Chlamydomonas reinhardtii* CW-15. *Arch. Microbiol.* **131**: 197–202.

258 Kuhlemeier, C., Green, P.J. & Chua, N.H. (1987) Regulation of gene expression in higher plants. *Annu. Rev. Plant Physiol.* **38**: 221–257.

Author Index

Subject Index